T0177568

How to Treat Persons

How to Treat Persons

Samuel J. Kerstein

OXFORD

UNIVERSITY PRESS

OXFORD

UNIVERSITY PRESS

Great Clarendon Street, Oxford, OX2 6DP,
United Kingdom

Oxford University Press is a department of the University of Oxford.
It furthers the University's objective of excellence in research, scholarship,
and education by publishing worldwide. Oxford is a registered trade mark of
Oxford University Press in the UK and in certain other countries

© Samuel Kerstein 2013

The moral rights of the author have been asserted

All rights reserved. No part of this publication may be reproduced, stored in
a retrieval system, or transmitted, in any form or by any means, without the
prior permission in writing of Oxford University Press, or as expressly permitted
by law, by licence or under terms agreed with the appropriate reprographics
rights organization. Enquiries concerning reproduction outside the scope of the
above should be sent to the Rights Department, Oxford University Press, at the
address above

You must not circulate this work in any other form
and you must impose this same condition on any acquirer

British Library Cataloguing in Publication Data

Data available

Library of Congress Cataloging in Publication Data

Data available

ISBN 978-0-19-969203-3

Printed and bound by CPI Group (UK) Ltd, Croydon, CR0 4YY

Links to third party websites are provided by Oxford in good faith and
for information only. Oxford disclaims any responsibility for the materials
contained in any third party website referenced in this work.

2

For Jo Ann and Howard Kerstein

Contents

Part II Practice

Acknowledgments

This book owes its existence to the generous support of people and institutions.

I would like to thank the University of Maryland, College Park, for granting me two semesters free of teaching duties to complete the book.

Many of the book's main ideas emerged when I was a Fellow in the Harvard University Program in Ethics and Health. I am very grateful to the program for its financial and intellectual backing. I had the privilege of launching into research in bioethics with the guidance of some of the world's foremost practitioners. I owe a debt of gratitude to Dan Brock, Norman Daniels, and Frances Kamm. I have also learned a great deal from Nir Eyal and Sadeth Sayeed. Without the help of Daniel Wikler, both philosophical and practical, my work on markets in organs would never have gotten off the ground. I have found inspiration in his thinking regarding each one of the bioethical issues addressed in the book.

Greg Bognar, my fellow Fellow at the Program in Ethics and Health, is co-author of a paper on which Chapter 6 is largely based. His collaboration has been invaluable to the book's treatment of issues surrounding the fair distribution of life-saving resources. I am fortunate to have had the chance to work with him.

I would like to thank the many colleagues who have given me comments on parts of my manuscript or precursors to it. In addition to each of the individuals mentioned above, I am grateful to: Rüdiger Bittner, Tom Christiano, I. Glenn Cohen, Richard Dean, Lane DesAutels, Katrien Devolder, Steve Emet, Ryan Fanselow, Thomas Hill, Jr., Iwao Hirose, Aaron Hoitink, Scott James, Paulus Kaufmann, Matt King, Nikolaus Knoepffler, David Lefkowitz, Luc Noël, Derek Parfit, Peter Schaber, Shlomi Segall, Oliver Sensen, Matthew Smith, Neema Sofaer, Alan Strudler, Mark Timmons, Chris Vogel, Robert Wachbroit, and David Wasserman. I am particularly grateful for comments on my manuscript as a whole sent to me by Greg Bognar, Thomas Pogge, and Alan Wertheimer.

I would like to express my gratitude to two anonymous referees for providing insightful and ameliorative comments on my manuscript. I would also like to thank Peter Momtchiloff at Oxford University Press for lining up such helpful referees and, of course, for his championing of the project. I very much appreciate his and his colleagues' help in getting this book into print.

Parts of some of my published papers have been incorporated into this book: Chapter 2 includes much of "Death, Dignity, and Respect," *Social Theory and Practice* 35 (2009), 505–30; Chapter 3 is an expanded and significantly revised version of "Treating Others Merely as Means," *Utilitas* 21 (2009), 163–80; Chapter 4 is based on "Treating Consenting Adults Merely as Means," *Oxford Studies in Normative Ethics*

1 (2011): 51–74; Chapter 7 uses material from "Kantian Condemnation of Commerce in Organs," *Kennedy Institute of Ethics Journal* 19 (2009), 147–69, and "Autonomy, Moral Constraints, and Markets in Kidneys," *Journal of Medicine and Philosophy* 34 (2009), 573–85; Chapter 8 is based on "Saving Lives and Respecting Persons," (co-authored with Greg Bognar) *Journal of Ethics and Social Philosophy* 5(2) (2010), 1–20. I acknowledge with appreciation the permission of the publishers to use material from these papers.

Without the love and support of my wife, Lisa Strong, and my two children, Eli and Evelyn, this book would surely not exist. I am especially grateful to Lisa for her generous willingness to undergo a two-year disruption in life and career so that I could pursue research in bioethics.

1

Introduction

This book takes its inspiration from Immanuel Kant's "Formula of Humanity," which commands that we treat persons never merely as means but always as ends in themselves.[1] The book has two main goals. It aims, first, to develop some ideas suggested by the Formula of Humanity into clear, plausible moral principles. It builds a new, detailed account of when a person treats another merely as a means, that is, "just uses" the other. It also offers a novel approach to the question of what it means for persons to be ends in themselves, that is, to have a dignity that demands respect. The book is not, however, a work in Kant scholarship. In its effort to develop plausible principles from ideas suggested by the Formula of Humanity it focuses neither on defending interpretations of Kant's text, nor on developing principles that the historical Kant would accept. The book's second main goal is to show how the Kantian principles it develops can shed light on pressing issues in bioethics. The book explores the question of how, morally speaking, scarce, life-saving resources such as flu vaccine ought to be distributed, the morality of markets in organs (e.g., kidneys), and the ethics of doing research on "anonymized" biological samples and of conducting placebo-controlled pharmaceutical trials in developing countries.

According to the Mere Means Principle, as we will call it, it is morally wrong for a person to treat another merely as a means. The book is predicated on the view that, as one philosopher has put it, the Mere Means Principle is "both very important and very hard to pin down."[2] It is important in that many of us are attracted to the notion that it serves as a moral constraint, that is, sets limits to what we may do, even in the service of promoting important goods such as saving lives. Appeals to the Mere Means Principle occur in many realms of ethical inquiry. For example, authors invoke the idea that research on human subjects, management of employees, and criminal punishment is morally impermissible when it involves treating persons merely as means.[3] Not surprisingly, the Mere Means

---

[1] Immanuel Kant, *Groundwork of the Metaphysics of Morals*, trans. Mary Gregor (Cambridge: Cambridge University Press, 1996), 429. I am referring to Preussische Akademie (vol. IV) pagination, which is included in the margins of the Gregor translation. I cite the *Groundwork* as GMS.

[2] Jonathan Glover, *Choosing Children: Genes, Disability, and Design* (Oxford: Oxford University Press, 2006), 65. Glover adds that "'I came to see that he was just using me' is a damning moral criticism."

[3] Robert Levine, "Respect for Children as Research Subjects," in *Lewis's Child and Adolescent Psychiatry*, ed. Andres Martin and Fred Volkmar (Philadelphia: Lippincott, Williams, and Wilkins, 2007), 140; H. L. Haywood, "Rotary Ethics," *The Rotarian* 1918, 277; R.A. Duff, *Trials and Punishments* (Cambridge: Cambridge University Press, 1986), 178–9.

Principle also plays a role in debates regarding terrorism, pornography, surrogate motherhood, and copyright law.[4] The principle even features in ethical inquiry into professional and collegiate sports, hip hop music, and (fictional) vampires.[5]

Although the Mere Means Principle is frequently invoked—indeed, it "has become a virtual mantra in bioethics"—its meaning has remained obscure.[6] The book joins a debate with contemporary ethicists on what the Mere Means Principle amounts to and on whether it is credible. The book tries to capture with some precision an intuitive notion of treating others merely as means, according to which it is plausible to think that doing so is typically morally wrong.[7] The plausibility of the principle, as specified in the book, will be confirmed if, as I hope, it yields insight into morally problematic features of (some) medical research and (some) commerce in organs.

Critics have charged that the concept of the dignity of persons is useless.[8] Without criteria that permit us to discern when persons' dignity is violated, the concept does, as the critics claim, remain "hopelessly vague."[9] But the book discusses two accounts of dignity that aim to provide such criteria. The first, an orthodox Kantian account, is reasonably determinate, but has normative implications that are problematic enough to warrant the development of a new, Kant-inspired account. While not a complete account, the new one aims to identify conditions under which it is plausible to say that a person has failed to honor another's dignity and thereby acted wrongly.

Some thinkers—one might call them dignity-deflationists—suggest that any ethical weight a notion of dignity possesses derives solely from its incorporating an idea of respect for persons commonly employed in medical ethics.[10] According to this idea, respect for persons requires that we not interfere with the choices of an autonomous person, unless those choices harm another.[11] In medical contexts, respect for persons requires getting their voluntary, informed consent before treating them or using them

---

[4] Claudia Card, *Confronting Evils: Terrorism, Torture, Genocide* (Cambridge: Cambridge University Press, 2010), 131; Robert Baird and Stuart Rosenbaum, *Pornography: Private Right or Public Menace?*, Rev. ed. (Amherst, NY: Prometheus Books, 1998), 103, 106; Ruth Macklin, *Surrogates and Other Mothers: The Debates Over Assisted Reproduction* (Philadelphia: Temple University Press, 1994), 64; Jeremy Waldron, "From Authors to Copiers: Individual Rights and Social Values in Intellectual Property," in *Intellectual Property Rights: Critical Concepts in Law*, ed. David Vaver (London: Routledge, 2006), 129.

[5] Adrian Walsh and Richard Giulianotti, *Ethics, Money, and Sport* (New York: Routledge, 2007), 65–76; Sarah McGrath and Lidet Tilahun, "Hip Hop and Philosophy: Rhyme 2 Reason," ed. Derrick Darby and Tommie Shelby (Chicago: Open Court, 2005), 144; Christopher Robichaud, "To Turn or Not to Turn: the Ethics of Making Vampires," in *True Blood and Philosophy: We Wanna Think Bad Things with You*, ed. George A. Dunn and Rebecca Housel (Hoboken, N.J.: Wiley, 2010), 9.

[6] Alan Wertheimer, *Rethinking the Ethics of Clinical Research: Widening the Lens* (New York: Oxford University Press, 2010), 47.

[7] So the book tries to respond to philosophers who are skeptical that the Mere Means Principle can be specified in a plausible way, including Nancy Davis, "Using Persons and Common Sense," *Ethics* 94 (1984).

[8] See Ruth Macklin, "Dignity Is a Useless Concept," *British Medical Journal* 327 (2003); and Steven Pinker, "The Stupidity of Dignity," *The New Republic* (2008).

[9] Macklin, "Dignity Is a Useless Concept," 1420.

[10] ibid. 1419; and Pinker, "The Stupidity of Dignity."

[11] See Pinker, "The Stupidity of Dignity" as well as F. Daniel Davis and President's Council on Bioethics (US), "Human Dignity and Respect for Persons: A Historical Perspective on Public Bioethics," in *Human*

in research, as well as protecting their confidentiality. But the book tries to reveal that sometimes one respects persons *in this sense* and yet fails to honor their dignity. One can (but does not necessarily) do this in research on human subjects as well as commerce in human organs. Moreover, it is plausible to think that such failure to honor human dignity is often morally wrong, all things considered. Far from being useless, a notion of dignity can bring an ethical perspective to which narrow principles of respect for persons are blind.

This chapter provides background for the book's development and application of Kantian normative principles. It explains and defends the book's methodology (1.2), sets forth the notion of a person the book employs (1.3), and explores briefly the relations of one of its main focuses, namely the idea of treating others merely as means, to other, related ideas (1.4). But before turning to this background, it is helpful to have in view a sketch of how the book unfolds.

1.1 Précis

The book is divided into two main parts. The first develops moral principles, while the second applies these principles to practical problems in bioethics.

Kant holds the Formula of Humanity to be (one formulation of) the supreme principle of morality. In his view, *all* moral duties derive ultimately from it. Moreover, not only is any action's moral permissibility (or requiredness) determinable through appeal to the Formula of Humanity, but no action that fails to accord with it can have any moral worth or goodness, in his view.[12] The book develops what many philosophers take to be important aspects of the Formula of Humanity: it offers accounts of treating others merely as means and of honoring the dignity of persons. But the book does not try to reconstruct the Formula of Humanity as a whole. For example, it does not explore conditions under which one treats *oneself* merely as a means.[13] And it does not defend a complete account of conditions under which we do or do not respect the dignity of persons. In short, the book does not purport to generate Kantian principles that can jointly serve as the supreme principle of morality.

Here is how the book unfolds. According to one prominent way of interpreting the Formula of Humanity (which we refer to as FH, for short), we treat humanity as an end in itself just in case our actions express proper respect for the unconditional and incomparable value humanity possesses, that is, for its dignity. In order to understand and appreciate Kant's principle, we *need not* focus on the prescription never to treat

Dignity and Bioethics: Essays Commissioned by the President's Council on Bioethics (Washington, D.C.: President's Council on Bioethics, 2008), 27.

[12] For discussion of Kant's concept of the supreme principle of morality, see Samuel Kerstein, *Kant's Search for the Supreme Principle of Morality* (Cambridge: Cambridge University Press, 2002), chapter 1.

[13] For interpretation of Kant's view of conditions under which one treats oneself merely as a means, see Samuel Kerstein, "Treating Oneself Merely as a Means," in Monika Betzler, ed. *Kant's Ethics of Virtues* (Berlin: Walter De Gruyter, 2008).

persons merely as means. But, Chapter 2 argues, if we employ this Respect-Expression Approach, FH has implausible implications. Sometimes an action is right even if it leads to a person's death. The chapter specifies cases in which killing in self-defense and sacrificing one's life for others are each morally permissible, or so many of us believe. But according to the Respect-Expression Approach, Kant's principle yields the conclusion that these actions are wrong. We therefore have grounds for skepticism regarding FH interpreted in this way.

The Mere Means Principle states that it is wrong to treat others merely as means. If the argument of Chapter 2 is convincing, then a leading approach to FH, an approach that de-emphasizes this principle, suffers from serious shortcomings. It thus makes sense to try other approaches to FH, or at least to elements of it, including one that attempts to develop the prescription not to treat others merely as means into a clear and plausible principle. Chapter 3 explores various attempts to set out a sufficient condition for treating others merely as a means and thus acting (*pro tanto*) wrongly. These attempts are all inspired, but not limited, by what Kant actually says in the *Groundwork of the Metaphysics of Morals*. At least three distinct ways of formulating such a condition might, with some plausibility, be gleaned from the *Groundwork*. One might hold that an agent treats another merely as a means if the other cannot share the end she is pursuing in using him. Or one might contend that she treats him merely as a means if he is unable to consent to her using him. One might interpret this inability to consent in two different ways, namely in terms of a lack of opportunity to consent or, rather, in terms of it being irrational to consent. Using materials suggested by the *Groundwork*, this chapter tries to develop a plausible sufficient condition for an agent's treating another merely as a means and thereby acting (*pro tanto*) wrongly, namely the Hybrid Account.

We can deepen our understanding of the Mere Means Principle by formulating a sufficient condition for an agent's using another, but *not* merely as a means. Following a suggestion by Robert Nozick, someone might propose that when an agent uses another, she does not use him merely as a means if he has given his voluntary, informed consent to her using him.[14] This account, which appeals to actual consent, seems to have the virtue of being simple and direct. An alternative is to invoke a notion of possible consent, according to which an agent can consent to being used only if he can avert this use by withholding his consent to it. A possible consent account holds that when an agent uses another, she does not use him merely as a means if it is reasonable for her to believe that he can consent to her use of him. Chapter 4 explores the plausibility of these accounts. It contends that an actual consent account suffers from shortcomings to which a possible consent account is immune. The former account has the unwelcome implication that certain ineffective or unnecessary attempts an agent makes at coercing or deceiving another to serve as a means to her ends do not amount to her just using the other. An actual consent account thus fails to realize its promise of

---

[14] Robert Nozick, *Anarchy, State, and Utopia* (New York: Basic Books, 1974), 31.

giving us a simple yet plausible way to capture a sufficient condition for an agent's using another, but not merely as a means. The account can be altered so that it no longer has the unwelcome implication in question, but the altered account differs little, with respect to both complexity and content, from a possible consent account.

The final chapter in Part I of the book culminates in a new account of the dignity of persons. Part of what constitutes their dignity is their having a status such that they ought not to be treated merely as means. So Chapter 5 begins by filling out our understanding of treating others merely as means. It crystallizes the various accounts of plausible sufficient conditions for treating another merely as a means developed in earlier chapters. And it specifies a necessary condition for an agent's treating another in this way. With the aim of solidifying our understanding of some of our accounts of treating others merely as means, the chapter then applies them to stylized cases involving transplant surgeons, runaway trolleys, and so forth. These applications aim to reinforce the accounts' plausibility.

The chapter then develops a Kant-Inspired Account of Dignity (KID). The account holds that dignity is a special status held by persons. An agent's treatment of a person respects his dignity only if it accords with this special status. The status is such that an agent ought not to use persons merely as means, but he ought to treat them as having unconditional, transcendent value: value that has no equivalent in any set of non-persons.[15] Moreover, the status of a person is such that an agent ought to treat another as having a value that does not diminish as a result of what she does or of the agent's relation to her, apart from some specified exceptions. Finally, the status of persons is such that they ought to be treated as having a value to be respected, rather than a value to be maximized by producing as many of them as possible. In order to clarify KID as well as to highlight its strengths, the chapter revisits the examples that, according to Chapter 2, undermine the credibility of a more traditional Kantian account of dignity.

Part II of the book, which moves from principles to practice, begins in Chapter 6 with an exploration of what it means to respect persons or their dignity in contexts in which scarce, life-saving resources must be distributed. Examples of such resources are organs for transplant, treatment/vaccine for flu, and beds in intensive care units. The chapter contends that two accounts of respect for persons (or their dignity), namely the Respect-Expression Approach to FH and an Equal Worth Account, have implausible implications regarding allocation decisions. For example, neither account would allow us to privilege in our distribution of resources the saving of a 20-year-old who would thrive for an additional sixty years over the saving of an 80-year-old who would thrive for an additional five years. Building on KID, the

---

[15] As will become evident in Chapter 5, the term "transcendent" is not meant to have orthodox Kantian or theological connotations. To say that something is transcendent in the sense intended here is not to say that it is beyond possible experience or that it is non-material.

chapter develops a new perspective on what respect for the dignity of persons demands in such contexts. The chapter proposes a novel way of balancing two plausible allocation principles, namely a principle of preserving the most persons and one of preserving the most "person years," that is, years of life during which one retains one's personhood, as opposed, for example, to being comatose. The chapter explores what respect for the dignity of persons demands in vexing scenarios, for example, one in which we must choose between saving one person for twenty years or five persons for two years each, or a scenario in which we must decide whether to treat hundreds of people who suffer from a relatively minor condition or do an expensive, life-saving procedure on one person.

A chronic shortage in organs for transplant (e.g., kidneys or lobes of liver) results in hundreds of deaths per year in the United States. In light of this shortage, some physicians and philosophers have championed the creation of a regulated market in organs, especially a market in which live "donors" may sell a kidney for profit. Chapter 7 explores conditions under which organ buying or selling would be incompatible with respect for the dignity of persons. The chapter sketches Kant's argument or, more precisely, a reconstruction of his argument, for condemning any sale by a person of one of her internal organs.[16] The chapter contends that this argument, which appeals to the Respect-Expression Approach to FH, fails. However, if we appeal instead to a different account of persons' dignity, namely KID, we find that the buying and selling of organs would often, but not always, fail to respect it.

The book's final chapter applies the Kantian notion of respect for the dignity of persons developed in the first part of the book to two ongoing controversies in research ethics. The first controversy can be illustrated with a hypothetical case. Suppose that a group of people with cystic fibrosis give their informed consent to investigators to have their blood used in a study of the effectiveness of mucus-thinning enzymes in preventing lung infections. After the investigators "anonymize" the blood samples, they provide them to an outside researcher who, without the patients' knowledge or consent, uses them for a study of a method for detecting whether early stage fetuses carry the mutation for cystic fibrosis. At least some of the patients would object to furthering such research, which might lead to an increase in the abortion rate of fetuses carrying the cystic fibrosis mutation. But neither the original investigators nor the outside researcher would, by virtue of the actions described, violate current United States federal regulations governing research. In cases such as this, it is nevertheless natural to wonder whether any of the experimenters treat their subjects merely as means and thus act (*pro tanto*) wrongly. Chapter 8 argues that the risk that they do so is significant and that there is therefore

---

[16] Immanuel Kant, *The Metaphysics of Morals*, trans. Mary Gregor (Cambridge: Cambridge University Press, 1996), 423. I am referring to Preussische Akademie (vol. VI) pagination, which is included in the margins of the Gregor translation. I cite *The Metaphysics of Morals* as MS.

reason to champion regulations that require consent for research on biospecimens, even if they have been anonymized.

A second controversy in research ethics addressed in the chapter surrounds the following sort of case. An American pharmaceutical company does a placebo-controlled drug trial in a resource-limited, developing country—a trial that stands to benefit both the company and the participants. But the trial would not be conducted in the United States, at least in part as a result of its failing to conform to ethical norms. If any trial of the drug were carried out in the US, the drug's effectiveness would not be measured against placebo, but rather against another drug: one belonging to the US standard of care. Bioethicists and philosophers are divided regarding the moral permissibility of conducting such a placebo-controlled trial. With the help of KID, the chapter investigates the moral permissibility of doing so.

Part I of the book, as well as the remaining sections of this chapter, contain detailed discussions in ethical theory—discussions necessary to provide a cogent rationale for developing new Kantian principles and to specify the principles with the clarity that renders them useful. But the discussions presuppose no expertise in Kant scholarship. They are peppered with concrete cases, some of which are drawn from bioethics and many of which are relevant to it. Bioethicists will, I venture, find Part I close enough to practical concerns to warrant their engagement and rigorous enough to be of genuine service to them. However, Part II is designed itself to give those who have minimal interest in ethical theory enough familiarity with the Kantian principles the book develops to appreciate their application to contemporary issues in bioethics.

Part II contains more appeals to facts on the ground than some ethical theorists might be accustomed to. Stylized cases regarding trolleys, criminally aggressive transplant surgeons, and so forth help to reveal strengths and weaknesses of moral principles, but so do cases that have greater verisimilitude. Application of principles to situations rich with real-world complexity expands our evidence regarding their plausibility. In particular, principles gain in credibility if they enable us to discern in actual (or close to actual) situations ethical features that might otherwise have been hidden to us.

1.2 Methodology

Although the book develops Kantian principles, it does not do so with the help of the kind of argument those familiar with Kantian ethics might expect. In Section III of the *Groundwork of the Metaphysics of Morals*, Kant tries to establish that the categorical imperative is valid: all rational agents are always bound, all things considered, to act in accordance with it (though some of us might fail to do so). This "deduction" of the categorical imperative, mentioned briefly in our discussion of Kant's notion of autonomy below, is particularly ambitious. Kant appears to take it to rest *solely* on premises that he proves to be irrational to deny, since, for example, denying them involves self-contradiction.[17] Kant himself seems to abandon the effort to construct this sort of foundational argument for the validity of the categorical imperative.[18] And this book makes no attempt to construct such an argument for any of the Kantian principles it embraces.

Kantian ethicists also construct arguments that are slightly less ambitious than this first type. These arguments try to establish that if one makes some initial, very general assumption about rational agency or morality, then one is rationally compelled to embrace some Kantian principle or value. But these "conditional foundational arguments," as we refer to them, do not purport to prove that denying this initial assumption is irrational.[19] A well-known example of a conditional foundational argument is to be found in Sections I–II of the *Groundwork*. One of Kant's aims in these sections is to prove that *if* we believe there to be a supreme principle of morality, then we are rationally compelled to hold it to be the Formula of Universal Law or some equivalent principle. (The Formula of Universal Law is the following: "*act only in accordance with that maxim through which you can at the same time will that it become a universal law.*"[20]) Kant's argument does not try to show it to be irrational to deny that there is a supreme principle of morality, that is, (roughly) an unconditionally and universally binding principle from which all genuine moral duties derive. The argument leaves open the possibility that we are justified in believing that there is no such principle.[21] Another example of this sort of argument is Christine Korsgaard's attempt to establish that if we assume that we act for reasons, we are rationally compelled to value for its own sake the humanity in ourselves and others.[22] This argument does not itself purport to show that it would be irrational to deny that we act for reasons.

After probing in detail these arguments and others like them, I have defended the conclusion that they fail.[23] Kant does not prove that assuming there to be a supreme

---

[17] But see Dieter Schönecker, "How is a Categorical Imperative Possible?" in Christoph Horn and Dieter Schönecker ed. *Groundwork for the Metaphysics of Morals* (Berlin: Walter de Gruyter, 2006).

[18] Immanuel Kant, *The Critique of Practical Reason*, trans. Mary Gregor (Cambridge: Cambridge University Press, 1996), 47. I am referring to Preussische Akademie (vol. V) pagination, which is included in the margins of the Gregor translation. I cite *The Critique of Practical Reason* as KpV.

[19] But advocates of the arguments might try elsewhere to show that denying this assumption is irrational.

[20] GMS 421.

[21] See, for example, GMS 425.

[22] Christine Korsgaard, *The Sources of Normativity* (Cambridge: Cambridge University Press, 1996), 90–130.

[23] For my criticism of Kant's "derivation" of the Formula of Universal Law, that is, of his attempt to prove that if we take there to be a supreme principle of morality, then we must conclude that it is this principle, see Samuel Kerstein, *Kant's Search for the Supreme Principle of Morality* (Cambridge: Cambridge University Press, 2002), 73–191. I try to refute Korsgaard's claim that assuming ourselves to have reasons for actions rationally commits us to holding humanity to be valuable in itself in Samuel Kerstein, "Korsgaard's Kantian Arguments for the Value of Humanity," *Canadian Journal of Philosophy* 31 (2001). In *Kant's Search*, I also attempt to expose gaps in a well-known reconstruction of Kant's argument for the claim that if we assume we have "transcendental" freedom, then we rationally commit ourselves to the Formula of Universal Law (33–45). And I criticize (46–72) Kantian attempts to show that if there is a supreme principle of morality, then it is the Formula of Humanity (a principle Kant held to be extensionally equivalent to the Formula of Universal Law). More recently (Kerstein, "Autonomy and Practical Law," *Philosophical Books* 49 (2008)), I have questioned an argument that Andrews Reath attributes to Kant (see Reath, *Agency and Autonomy in*

principle of morality rationally compels us to embrace any principle he champions as such, and Korsgaard does not show that taking ourselves to have reasons for our actions entails that we must, rationally speaking, value humanity for its own sake. I find unconvincing all of the defenses I have encountered of conditional claims asserting that if we embrace some very general proposition regarding morality or rational agency, Kantian principles or values become rationally irresistible. And since I am not sanguine about the prospects of defending such a claim, I do not try to do so here.

But I do offer a basis for thinking ourselves justified in embracing the Kantian principles developed in the book. I try to show that, according to many of us, they do a better job than many other Kantian principles in yielding prescriptions that cohere with the dictates of our reflective common sense, that is, with our considered moral judgments or, equivalently, intuitions.[24] For example, I develop a sufficient condition for an agent's treating another merely as a means and thereby acting (*pro tanto*) wrongly. I argue that, according to many of us, this Hybrid Account has overall plausible implications regarding when an agent's use of another is wrong, whereas several other accounts one might propose do not. But I do not treat considered moral judgments as sacrosanct. For example, I acknowledge that the account I develop of using another, but not merely as a means, clashes with a judgment many of us are inclined to make, namely that a surgeon who overcharges a patient for a life-saving operation is just using him. But I suggest that the power of this account to capture our judgments in a wide range of cases gives us good reason to give up this particular judgment. In sum, I try to show that we have some justification for embracing the Kantian principles I develop. This justification lies in their ability to generate moral prescriptions acceptable to our reflective common sense in a variety of familiar contexts, as well as unfamiliar ones, such as that of commerce in organs.

But this justification is far from complete. First, a more thorough defense of the principles would involve considering how they fare against a wide range of competing principles or ethical views. For example, some of the Kantian principles developed here are moral constraints: they imply that it can be wrong to do something (e.g., treat another merely as a means), even if doing it would have good consequences overall. A more complete defense of the principles would consider their plausibility relative to consequentialist views that reject moral constraints. A consequentialist might, of course, argue that when we consider the whole range of our considered moral judgments, we find that we are committed to the view that the rightness of actions

---

Kant's Moral Theory (Oxford: Oxford University Press), 2006, chapter 5), according to which assuming that we have autonomy in one particular Kantian sense rationally compels us to view the Formula of Universal Law to be binding on us.

[24] In the usage I adopt, considered moral judgments or intuitions can be either judgments regarding particular cases or kinds of cases (e.g., "It was morally permissible for that soldier to have sacrificed his own life in order to save the lives of the others riding in the tank with him"; "It is morally permissible for someone to take his own life in order to save the lives of others") or judgments of a more general nature (e.g., "It is wrong to treat another merely as a means").

depends solely on the goodness of their effects. To take another example, the book does not weigh the plausibility of embracing the Kantian principles it develops against "particularist" views, according to which it is a mistake to think that moral judgment involves the application of *any* general principles to cases. And the book does not confront the skeptical view that no one has any moral obligation to do anything at all. I do not believe that either moral particularism or moral skepticism is so well supported that it undermines the credibility of Kantian principles. But this book does not undertake to support this belief.

The book's limited defense of Kantian moral principles uses a method akin to that of reflective equilibrium, suggested by John Rawls.[25] The book tries to arrive at Kantian principles that cohere with a wide range of our considered moral judgments, and do so better than rival Kantian principles. It tries to put plausible Kantian principles forward for further discussion and evaluation. As just suggested, this further evaluation would involve weighing evidence for these Kantian principles against that for a range of opposing moral principles. It would also include measuring evidence for views on the nature of morality that are consistent with the legitimacy of these Kantian principles against views that would undermine their legitimacy. Further defense of the principles would involve employing something like the method of *wide* reflective equilibrium, according to which one tries not only to bring moral principles and considered moral judgments into equilibrium with one another, but also with background theories regarding the nature of morality, human psychology, and so forth.[26]

This general approach to justifying one's acceptance of a moral theory has attracted lively criticism. According to Allen Wood, it is superficial. It aims only to systematize our beliefs, leaving them "without any firm foundation."[27] Moreover, since this method aims at coherence among extant beliefs, it provides little basis for significant revisions in them.[28] Wood contrasts this approach to justifying ethical principles and values with what he calls the "philosophical" model.[29] This model sees the fundamental principle in ethical theory as resting on a fundamental value. To ground adequately this fundamental value is, according to this model, to show that we are committed to it "simply in rationally desiring ends and willing actions toward them."[30]

In order to assess Wood's criticism of the method of reflective equilibrium, it will be helpful to see at work the method he prefers to it. According to Wood, "Kantian ethics rests on a single fundamental value—the dignity or absolute worth of rational nature, as giving moral laws and as setting rational ends."[31] Wood tries to defend the claim that

[25] John Rawls, *A Theory of Justice* (Cambridge, Mass.: Harvard University Press, 1971), 48–53.

[26] For helpful development of a notion of wide reflective equilibrium, see Norman Daniels, "Wide Reflective Equilibrium and Theory Acceptance in Ethics," *Journal of Philosophy* 76 (1979): 256–82; and Ryan Fanselow, "Moral Intuitions and Their Role in Justification." PhD Dissertation. University of Maryland, College Park, 2011.

[27] Allen Wood, *Kantian Ethics* (Cambridge: Cambridge University Press, 2008), 51.

[28] Wood, *Kantian Ethics*, 60, 65.

[29] ibid. 54.   [30] ibid. 55.

[31] ibid. 94. See also 257.

persons have such worth.[32] He begins by setting forth a Kantian assumption: in setting an end, that is (apparently), in choosing it and trying to realize it, you commit yourself, rationally speaking, to adopting means you view to be necessary to realize it. So, for example, if you are trying to lose weight and believe that in order to do so you need to avoid desserts, then you are, rationally speaking, committed to avoiding them, regardless of how attracted you are to caramel tarts, marzipan, and so forth. But if you think of the means to your end as good, then you must also think of your end as good, Wood's argument continues. Moreover, you must regard your own rational capacities "as authoritative for what is good in general," for you treat these capacities as capable of determining which ends are good.[33] But in thinking of your own rational capacities as authoritative in this way, you also esteem yourself as unconditionally good. So, in short, if you take yourself rationally to set ends, then you are committed to the unconditional value of your rational nature. As Wood is well aware, a further step is needed to show that you are committed to the unconditional value of everyone's rational nature. But let us just assume that if Wood's argument establishes that each of us must think of himself as having such value, then he must also think of all other persons as having it as well.

This argument seems to exemplify the conditional foundationalist approach. It apparently tries to show that if one makes some initial assumption about rational agency, in this case an assumption about what, rationally speaking, is entailed by setting an end, then irrationality is the price of refusing to embrace a Kantian claim regarding the special worth of persons. Like other arguments that take this approach, this one suffers from significant gaps, or so I try to show. If I am correct, there is no reason to think that the argument would provide the Kantian conclusion regarding the value of humanity a stronger foundation than would one that appealed, among other things, to the degree to which the conclusion cohered with our considered moral judgments.

According to a key step in the argument, your thinking that you have the capacity to determine which ends are good commits you to the view that you yourself are good. But your thinking this does not commit you to that view any more than your thinking that you have the capacity to determine which ends are bad commits you to the view that you yourself are bad. Neither in general nor, so far as I can tell, in this case, does your being committed to holding that you have a capacity to determine whether something has a certain property rationally compel you to conclude that you possess that property.

Moreover, even if it succeeded, the argument would show that our making Wood's Kantian assumption would rationally commit us to viewing persons as unconditionally valuable in the sense of valuable no matter what the context of their existence is or might be. But, according to Wood, another aspect of the value of persons, which is the

---

[32] Wood's account of persons differs slightly from the one developed here, but the differences do not affect the discussion that follows.

[33] Wood, *Kantian Ethics*, 91.

fundamental value of his Kantian ethics and the value his argument presumably aims to show us we are committed to embracing, is their dignity. A being with dignity "has a value that may not be rationally traded away or sacrificed, not even for something else that has dignity," Wood says.[34] Yet that some being, say a person, is unconditionally valuable does not imply that it is never legitimate to trade her away for the sake of preserving other unconditionally valuable beings, say five other persons. That there can be no context in which a person exists yet has no value fails to entail that the value of persons does not aggregate. It fails to entail that it is inappropriate to think of five persons as having *more* value than one person and *on that basis*, for example, to save the five rather than the one in a tragic situation in which one cannot do both. Let us even suppose that Wood's argument demonstrates persons to have unconditional value in the following sense: not only do they possess a positive worth in whatever context they do or might exist, but each individual person has no more or no less worth than any other individual.[35] Still we are left with the possibility that the worth of persons is aggregative and thus that it can be legitimate to trade away that inherent in some persons in order to secure that inherent in (a greater number of) others. Wood provides nothing to bridge the gap between establishing that persons have unconditional value and establishing that they have dignity.[36] Even if we make Wood's initial Kantian assumption regarding the rational commitment entailed by setting an end, the philosophical method, as he employs it, falls far short of providing secure grounds for the conclusion that humanity has the special value his Kantian ethics attributes to it. Wood's argument does not give us a basis for a significant revision of our moral views.

In the end, Wood suggests that his argument is less aspiring than we have construed it. He seems to acknowledge that it fails to prove that if we make his Kantian assumption about setting ends, we are rationally compelled to hold humanity to have the special value he attributes to it. Rational action, he says, can be conceived of in ways that do not commit oneself to holding humanity to have such value. But, he claims, the "interpretation" he advocates (which is encapsulated in the argument we summarized) is more "natural and reasonable" than other interpretations of it.[37] Wood says very little about what these alternatives are or about why, precisely, the Kantian interpretation has an advantage over them.[38] In any case, his claim prompts a question:

---

[34] ibid. 94, 180.

[35] Wood suggests that for Kant (and presumably for Wood as well) to say that a being has absolute value is in one sense just to say that it has unconditional value, as just described. But in another sense for something to have absolute value is for it to have dignity. See ibid. 94 and 291, note 6.

[36] Wood says simply that in "combining [the Formula of Universal Law] with [the Formula of Humanity] and advancing to [the Formula of Autonomy], Kant makes a further claim about the moral status of rational nature in persons: He claims that it has 'dignity' (*Würde*)" (ibid. 94).

[37] ibid. 93.

[38] Wood says that "the representation of something as an end might be taken as a merely theoretical act of perceiving the goodness of an object, a passive state that would move us of itself, rather than an act of rational judgment carrying with it a practical authority for us that is worthy of esteem as an end in itself" (ibid. 93).

what does it mean for a conception of rational action to be more "natural and reasonable" than another if not that it coheres better than the other with our considered beliefs? It would be odd, to say the least, to think that conceptions of rational action come with degrees of naturalness or reasonableness built into them, independently of how they relate with other views we have of ourselves and the world. In light of the gaps in Wood's argument, I venture that the naturalness and reasonableness that some might see in it is largely a function of its conclusion cohering well with their considered judgments—including their considered moral judgments. In the end, Wood seems to resort to something like the method of wide reflective equilibrium in order to justify his endorsement of the notion that humanity has unconditional worth and dignity. The philosophical method, as he employs it, does not constitute a genuine alternative to it.

Moreover, the method of wide reflective equilibrium is compatible both with the project of giving Kantian principles a conditional foundation and with the even more ambitious one of establishing them through premises one shows it to be irrational to deny.[39] If the latter project is successful, then it presumably gives us strong reason to accept Kantian principles and to abandon any moral judgments that conflict with them. If the former is successful, then it might also give us strong reason to abandon recalcitrant convictions, depending on the cost that embracing the condition would itself have in terms of its coherence with our other considered judgments. Of course, if either of these projects comes to fruition, then some people, for example, act consequentialists, might acquire a basis for making significant revisions in their moral beliefs. Contrary to Wood's charge, the method of wide reflective equilibrium is not inherently conservative. I refrain from joining in these foundational projects, not because I think that they clash with sound methodology, but rather because I do not know how to contribute to the prospects of their success, which I do not believe to be very good.

Another objection to the method employed in this book concerns its reliance on considered moral judgments. The book appeals to such judgments or, equivalently, intuitions regarding the moral valence of actions as well as regarding more general moral propositions. It appeals, for example, to the intuition that in certain circumstances, it is morally permissible for someone intentionally to kill another in self-defense as well as to the intuition that persons have a special status such that using them in some ways is morally wrong, even if doing so produces the best consequences. These intuitions play a role in the process of progressing towards a justification of moral principles, for this process involves selecting principles that belong to a coherent whole including the principles themselves, intuitions, and background theories. Intuitions might influence not only which moral principles, but even perhaps which background theories we accept. But Peter Singer suggests that our intuitions, or at least some of them, do not merit our giving them such a role. For they "are likely to derive from

---

[39] I owe this point to Ryan Fanselow.

discarded religious systems, from warped views of sex and bodily functions, or from customs necessary for the survival of the group in social and economic circumstances that now lie in the distant past."[40]

In response, let us suppose that we trace the genealogy of some intuition, say that masturbation is morally wrong, and find that it is based on a religious system we disavow. Our disavowal of that religious system would presumably be supported by one of our background theories. Since we are seeking to reach equilibrium not only between moral intuitions and principles, but also between these and our background theories, we would have reason to refrain from counting against a moral principle we are considering adopting that it entailed there to be nothing wrong with masturbation. Moreover, we would have reason not to weigh in the principle's favor that it entailed masturbation to be wrong. The general point is that since our method does not treat intuitions as fixed points, it assigns a limited justificatory role to them. Application of the method could even result in our abandoning all particular intuitions regarding right or wrong. That would happen if it turned out that a background theory entailing moral skepticism had especially strong support. Again, the method of wide reflective equilibrium might lead us to significant changes in our moral views.

Citing recent work by experimental psychologists, Singer suggests that some deontological intuitions—for example, the intuition that it is wrong to push someone off a bridge and onto the tracks in front of an on-coming trolley, even if that is the only way to prevent it from killing five others—are based on a certain kind of emotional processing in the brain. And this processing responds to morally irrelevant factors such as that saving the five in this case involves a close-up kind of violence rather than morally relevant ones, such as that one person dying is better than five dying. By contrast, consequentialist intuitions—for example, the intuition that in this trolley case it is not wrong to push the one onto the tracks—are not based on this sort of emotional processing. And they respond to morally relevant factors such as the one just mentioned. So, Singer suggests, in the justification of moral principles, deontological intuitions ought to carry no weight.[41]

If this argument is persuasive, then, according to the method of wide reflective equilibrium, we would have good reason to discount deontological intuitions. The method would be responsive to such findings. But Singer claims that responsiveness of this sort entails that the method is "vacuous." According to him, "the "data" that a sound moral theory is supposed to match have become so changeable that they can play, at best, a minor role in determining the final shape of the normative moral theory."[42] This claim is puzzling. That "data," including considered moral judgments about general rules or particular cases, are changeable, for example, such that they can

---

[40] Peter Singer, "Sidgwick and Reflective Equilibrium," *The Monist* 58 (1974): 516. See also Singer, "Ethics and Intuitions," *The Journal of Ethics* 9 (2005): 348.

[41] ibid. 347–8.

[42] ibid. 349.

be discredited, does not entail that they *will change*, for example, that they *will be* discredited. Whether intuitions get discredited depends, in part, on the background theories that turn out to be best supported. If our best supported background theories do not undermine our intuitions, then the latter will have a significant role in determining the shape of the normative principles we accept. (Of course, our moral intuitions would presumably play some role in our deciding which background theories to accept as well.)

In any case, Singer's argument is not convincing. He does not show that deontological judgments respond to morally irrelevant aspects of situations.[43] First, Singer does not establish which aspects of situations those who make deontological judgments are responding to. For example, he suggests that in the trolley case described they are responding negatively to the close, interpersonal violence involved in pushing one person off of a bridge to save five others. But he does not eliminate other possibilities, for example that at least many of them are responding negatively because they believe that killing the one to save the others would amount to "just using" him.

Second, Singer's judgments that certain factors of situations are morally relevant or irrelevant are normative. They do not derive from what scans reveal about the parts of the brain that are active before one has a certain intuition. But what is the basis of these judgments, if not Singer's own intuitions? If these intuitions are consequentialist, privileging the importance of consequences over the importance of how agents act in order to bring them about, then he would seem to be begging the question against deontology.

Perhaps Singer would deny that his judgments regarding moral relevancy are based on intuitions in the sense of ordinary considered moral judgments about particular cases and insist instead that they are based on self-evident fundamental axioms. But if an appeal to Singer's own intuitions is in danger of being question begging, then so is an appeal to (allegedly) self-evident fundamental axioms. As an example of a self-evident axiom, Singer suggests Henry Sidgwick's notion that "the good of any one individual is of no more importance, from the point of view . . . of the universe, than the good of any other."[44] But not only do many Kantians fail to find this self-evident, they believe it to be false, at least if it is interpreted to be saying that the well-being of any one is of no more importance from this point of view than the well-being of any other. For, according to these Kantians, an impartial spectator (presumably taking the point of view of the universe) would hold it to be far more important to ensure the happiness of a good person than that of an evil one. And some Kantians make their own appeals to (something like) a self-evident axiom, namely that the categorical imperative is binding on all rational

[43] This discussion has been influenced by Selim Berker, "The Normative Insignificance of Neuroscience," *Philosophy and Public Affairs* 37 (2009): 321–7. See also Richard Dean, "Does Neuroscience Undermine Deontological Theory?" *Neuroethics* 3 (2009): 43–60.
[44] Henry Sidgwick, *The Methods of Ethics*, 7th ed. (Indianapolis, Indiana: Hackett, 1981), 382.

agents. They follow Kant in holding that "the moral law is given . . . as a fact of pure reason of which we are a priori conscious and which is apodictically certain."[45] I do not wish to endorse this claim, but rather simply to emphasize that what is to one philosopher a self-evident axiom is to another a path to moral error. And Singer offers us no guidance in deciding which is which.

A final point regarding methodology is in order. Kant holds that it is always wrong for a person to treat another merely as a means or to fail to honor someone's dignity. In his view, there are no possible circumstances in which doing these things is morally permissible. As it might already be apparent, the book does not try to defend that view. It holds that it is wrong *pro tanto* to treat another merely as a means or to fail to honor his dignity, but that, depending on the circumstances, acting in these ways might not be wrong all things considered. The claim that it is wrong *pro tanto* to act in these ways should be understood to mean that we have strong reasons not to do so. But these reasons might be outweighed by other reasons. For example, suppose that the only way to save millions of people who are threatened with death in a nuclear explosion is to kill one innocent person as a means to preventing it, thereby treating him merely as a means. We have a weighty reason not to treat the innocent person in this way, but this reason is outweighed by our reasons to save the millions. So the notion that it is wrong to treat others merely as means places a defeasible constraint on our action, according to the view presented here.

The language of an action's being wrong *pro tanto* is sometimes used interchangeably with that of our having a *prima facie* duty not to do it. We will not adopt this practice. That one has a *prima facie* duty to refrain from performing an action can be taken to mean that one is, all things considered, obligated to refrain from doing so, unless one has a stronger *prima facie* duty to perform the action (in which case one has an all things considered duty to perform it). But we do not wish to commit ourselves to this sort of view, which might naturally be taken to imply that the *only* reasons strong enough to outweigh those not to treat others merely as means or to honor their dignity amount to ones that generate a duty to treat others merely as means or to violate their dignity. Let us leave open the possibility, for example, that it is sometimes legitimate to fail to honor someone's dignity even if we do not have an all things considered duty to fail to honor it. Circumstances in which such a possibility might be realized are discussed in Chapter 5.

1.3 Persons

The book employs the term "person" in a quasi-technical, Kantian sense. The use of "person" is quasi-technical in that, as will soon become evident, the term refers neither to all human beings nor, necessarily, to human beings alone. The use of "person" is

---

[45] KpV 47.

Kantian in that it reflects what some contemporary Kantians consider (plausibly, in my view) to be central elements of Kant's notion of a person.[46] But I claim neither that Kant nor any of his contemporary interpreters uses the term precisely as I do. Let me also make clear from the outset that the concept of a person employed here is a threshold concept. If one has the set of capacities that are constitutive of it, one has personhood, no matter how well- or ill-developed those capacities may be.

"Person" refers here to all and only beings who have rational nature or, equivalently in the book's usage, "humanity." To have humanity is to have certain capacities. First, if a being has humanity, then she is an agent. She has the capacity to set and pursue ends.[47] She can represent to herself some outcome (e.g., that of a painting's being preserved) and try to bring it about (e.g., by putting the painting into a climate-controlled room). The capacity to set and pursue ends is fundamental to being a person. It is evident that if one does not have this capacity, then one does not have the others we are about to describe, with one possible exception mentioned below.

Second, a being who has humanity can conform what she does to practical rules that she has in view. By practical rules (or simply rules, for short) I mean rules that prescribe action. First, she can set for herself and abide (or fail to abide) by rules for her own action, for example, "I will exercise during my free time in order to stay in shape." Kant calls such rules "maxims."[48] A being who has humanity can also act with the help of rules that have a wider scope, indeed ones that apply to all persons. For example, she might guide her effort to solve a math problem with a rule such as: "If you are trying to determine what constitutes 30 per cent of 500, then you ought to multiply 500 by .3."[49] Kant calls rules like this one, namely ones that specify means required to attain some end we are seeking or might seek, "hypothetical imperatives." That persons have the capacity to direct their pursuit of ends with such imperatives in mind entails that they have the capacity to pursue ends rationally.

Kantian persons not only have the capacity to follow rules that specify means to their ends but the related capacity to try to ensure that their ends fit together coherently and efficiently. For example, they can decide not to set as an end securing something they want now, such as a new car, in order to realize some end in the future, such as owning a home. Persons can not only pursue ends for immediate satisfaction, but pursue

---

[46] In constructing this Kantian notion of a person, I have been influenced by Thomas Hill, *Dignity and Practical Reason in Kant's Moral Theory* (Ithaca: Cornell University Press, 1992), 38–41; Allen Wood, *Kant's Ethical Thought* (Cambridge: Cambridge University Press, 1999), 118–20; and Richard Dean, *The Value of Humanity in Kant's Moral Theory* (Oxford: Clarendon Press, 2006), 24–33. Wood and Dean imply that they would balk at attributing to Kant the notion of a person sketched here, and it is by no means clear that Hill would attribute it to him. Of course, it is not my intention to attribute this view to Kant.

[47] Kant tells us that "the capacity to set oneself an end—any end whatsoever—is what characterizes humanity (as distinguished from animality)" (MS 392).

[48] See GMS 421. Although on my account persons can act on maxims, I do not endorse Kant's view that each and every action a person performs is done on some maxim.

[49] I interpret hypothetical imperatives to have a wide scope. For example, this rule applies to all persons, not just to ones who have the end of determining what constitutes 30 per cent of 500, in that it specifies to all persons that if they will this end, then they ought to do something.

projects that they believe will benefit them in the longer term. But it is important to keep in mind that we are employing a threshold concept of a person. Given that he has pursued some ends rationally and can do so again, a dreamer who is now drifting from one project to another is not thereby any less (or more) of a person than a decision theorist who has a maximally coherent set of ends and pursues them with utmost efficiency.

It is perhaps already implicit in the capacities we have attributed to persons that, as beings with rational nature, they are not only able to guide their actions with rules specifying what ought to be the case, but also to recognize what is the case. A person who wants to visit Grant's tomb cannot really guide himself by the rule "In order to visit Grant's tomb, you ought to travel to New York" unless he has some under-standing of what a city and a tomb are. In any case, persons have the capacity to perceive and to understand what is in their environment. (Perhaps there could be a purely contemplative being who has this capacity, but not the capacity to set and pursue ends. If so, having the latter capacity is not necessary for possessing all of the capacities constitutive of personhood.)

It also seems implicit in the capacities we have attributed to persons that they have the capacity of self-awareness. A being cannot give herself and abide by a rule for her own acting (e.g., a New Year's Resolution), unless she distinguishes between herself and other beings in the world and cognizes herself as a being that endures through changes over time. Take, for example, the rule "I will read *In Remembrance of Things Past* by the end of the year." In order to guide her action with this rule, a being would have to distinguish between herself and the world apart from her (e.g., between *her* reading the novel and someone else doing so). A being would also have to have some sense of herself enduring through time. She would have to recognize that the same she who started the book a few months ago is now about to finish it. Whether or not it is indeed implicit in persons' having the capacities already described that they also have self-awareness, on the Kantian notion persons have such awareness.

On my understanding, if a being is a person she also has the capacity abide by a rule that, in her view, specifies what persons within its scope ought to do unconditionally, that is, regardless of whether they believe that it would give them pleasure, satisfy some desire of theirs (other than that of conforming to the rule), or make them happy.[50] She has the capacity to abide by such a rule even if she does not believe that her doing so would give her pleasure, satisfy some desire of hers, and so forth. So being a person involves having the capacity to resist immediate temptations, both for the sake of (what one takes to be) satisfaction in the future and for the sake of (what one takes to be) moral reasons. According to Kant, rules that specify what everyone genuinely ought,

---

[50] Such rules, which many of us believe to specify moral requirements, are categorical imperatives in a broad sense of the notion. For an account of the various ways in which Kant uses "categorical imperative," see Samuel Kerstein, "Imperatives: Categorical and Hypothetical," *International Encyclopedia of Ethics*, ed. Hugh LaFollette (Oxford: Blackwell), 2012.

unconditionally, to do are "categorical imperatives" in one sense of the term. These rules specify moral requirements, he believes. Suppose a being considers the following to be such an imperative: "Never try to humiliate another." If that being is a person, then she has the capacity to conform her action to that rule. That does not mean, of course, that she will always do so. She might find herself in circumstances in which her humiliating another would win her adulation from peers and, for that reason, she might go ahead and humiliate someone. But being a person involves having the capacity to do what one takes to be morally necessary, even if one would gain more satisfaction from doing otherwise. Let us call this capacity "limited moral agency." We call it limited to avoid the impression that, in our view, it is the only capacity that contributes to making one a morally excellent person. It seems that other capacities also contribute to this, for example, the capacity to discern, with the help of imagination and feeling, what others need and how one can help them get it.

Limited moral agency is a different, seemingly more modest, capacity than one Kant himself attributes to persons, namely autonomy. (So there is a second sense in which limited moral agency is limited.) In the *Groundwork of the Metaphysics of Morals*, Kant tries to prove that a certain principle, namely the Formula of Universal Law, is not only a categorical imperative, but also the supreme principle of morality from which all genuine moral duties derive. His attempt to prove this is notoriously difficult to interpret. But along the way he tries to show that we must take our wills to be free, that is, able to bring about effects "independently of alien causes determining" them.[51] Yet viewing ourselves as free in this sense requires us to take a "standpoint" outside of the world of sense, that is, outside of the realm in which our wills are determined by alien causes, including desires and inclinations. In addition to the world of sense, we must think of ourselves as belonging to the "world of understanding."[52] This is a world in which we have freedom in a positive sense or, equivalently, autonomy. Autonomy is the capacity of our "proper" self, without being caused to do so by any preceding event, to set forth a law for our willing and to determine whether to conform our actions to it. Kant identifies this law as the Formula of Universal Law (or some close variant of it).[53]

In conceiving persons as beings with limited moral agency, we do not seem to be committing ourselves to the view that they have autonomy in this sense. To have limited moral agency is to have the capacity to abide by what one considers to be a categorical imperative, even if one does not believe that doing so will give her pleasure, satisfy one of her desires (other than that of conforming to the rule), or make her happy. But one might, it appears, have this capacity yet not the capacity to act spontaneously, that is, with no preceding event causing one's action. It might be, for example, that human persons are so constituted that reflecting on the idea that some

---

[51] GMS 446, italics removed.
[52] ibid. 451–2.
[53] ibid. 451–3, 460–1.

action is required by what they take to be a categorical imperative sometimes causes them to abide by it, even if they do not believe that they will thereby get pleasure or happiness. In order to serve as such a cause, an agent's reflection on this idea need not, of course, be spontaneous.

Moreover, in exercising limited moral agency, one might represent a principle other than the Formula of Universal Law as a categorical imperative. One might, for example, hold to be such an imperative a principle of utility, according to which persons ought always to do what they believe will maximize overall well-being. One might conform one's action to that principle, which presumably condones actions that violate the Formula of Universal Law, even though one does not believe that doing so will promote her own interests. Kant seems to hold that the Formula of Universal Law condemns the making of false promises. But we can imagine a politician who makes a false promise in order to promote overall well-being, even though she realizes that doing so will irreparably harm her reputation. Persons can exercise limited moral agency even in performing actions that Kant would hold to be morally wrong.

Let us put Kant's particular notion of autonomy aside.[54] There are other, contemporary notions of autonomy or, more precisely, of acting autonomously that are often associated with persons. It turns out that, according to two distinct contemporary notions of acting autonomously, persons as we conceive of them here are indeed beings who can, under appropriate conditions, do so.

According to a notion of autonomous action prominent in bioethics, someone acts autonomously if she acts intentionally, with understanding, and free of certain kinds of external or internal control.[55] Since this account is "specifically designed to be coherent with the premise that the everyday choices of generally competent persons are autonomous," let us call it the "everyday account."[56] For someone's behavior to amount to an intentional action, it must correspond to her conception of it. If she has no mental representation of her behavior, or a highly inaccurate one, then it is not intentional. Someone acts with understanding only if she has at least a basic grasp on the nature and consequences of what she is doing. If a patient believes that after she donates a kidney another one will grow back in its place, then she does not understand what she is doing. Persons in our sense can perform intentional actions. The capacity to pursue ends rationally obviously involves the capacity to have an idea of what one is doing in order to secure the end. Depending on circumstances, persons can obviously understand what they are doing and what its effects will be. And, depending on circumstances, they can also act free of the sort of control that, according to this account, would render their actions less than autonomous. They can, for example,

---

[54] For a sympathetic presentation of Kant's view of autonomy, see Onora O'Neill, "Autonomy: The Emperor's New Clothes," *Aristotelian Society Supplementary Volume* 77 (2003).

[55] Tom Beauchamp, "Autonomy and Consent," in *The Ethics of Consent: Theory and Practice*, ed. Franklin Miller and Alan Wertheimer (Oxford: Oxford University Press, 2010), 65. See also Tom Beauchamp and James Childress, *Principles of Biomedical Ethics*, 6th ed. (New York: Oxford University Press, 2009), ch. 4.

[56] Beauchamp, "Autonomy and Consent," 65. The label for the view of autonomy is mine.

act free of coercion by another (external control) and free of obsessive desires symptomatic of severe mental illness (internal control).

A second conception of autonomous action takes shape against the background of a prominent contemporary notion of what it means to have autonomy. According to this "split-level" notion, to have autonomy is to have the capacity to try to control or to identify with one's first-order preferences by willing to abide by or reflecting on one's higher-level preferences.[57] Someone might, for example, have a preference to fill his free time by playing video games. He just loves to play. But he might also have a preference not to be the sort of person who acts on this preference. The latter preference is higher-level in that it is about his other (first-order) preference. By virtue of having autonomy, the person would be able to (try to) refrain from satisfying his preference to fill his free time in this way.[58] It is evident that a person in the Kantian sense would have autonomy as just described. For a person can (try to) regulate her own conduct based on self-given rules, including rules that specify that she not try to satisfy some of her first-order preferences. Based on reflection regarding what sort of person he aims to be, our gamer might, for example, attempt to abide by a rule according to which he must spend some time outdoors before he can play video games.

Building on this notion of autonomy, one might say that a person's action is autonomous if and only if she is acting on some preference of hers and, based on reflection on her values, she either does, or, if she thought about it, would choose to have this preference even in light of understanding how it arose in her.[59] So, for example, our gamer's playing for yet another hour is autonomous if he is doing it because he enjoys it and, upon reflection, he would choose to do it, even with the knowledge that his love for virtual reality grew out of his social isolation as an adolescent. Since they have autonomy, Kantian persons obviously act autonomously if, in acting, they fulfill the conditions just described.

To say that Kantian persons can, under certain conditions, act autonomously according to the two notions we have briefly sketched is not to say that the two notions are the same, of course. One might act autonomously according to the one, but not the other. Suppose, for example, that our gamer would not choose to have the preference to spend his free time in virtual reality if he knew that the preference was born of his isolation as an adolescent. His devoting his free time to gaming would then not be an autonomous action, according to the split-level theory sketched here. But it

[57] See Beauchamp, 63; I borrow the label from him. A main inspiration for this notion of autonomy is to be found in Harry Frankfurt, "Freedom of the Will and the Concept of a Person," *Journal of Philosophy* 68 1971: 5–20. The notion of is, of course, specified differently by different philosophers.

[58] Of course, we can also imagine the person *without* the second-order preference not to be such a devotee of video games. By virtue of having autonomy, he might reflect and determine that his values (based on which he formulates his higher level preferences) are consistent with his trying to satisfy his first-order preference for filling his free time with gaming.

[59] For discussion that motivates and develops a more elaborate account along these general lines, see David DeGrazia, *Human Identity and Bioethics* (Cambridge: Cambridge University Press, 2005), 95–106.

would be autonomous according to the everyday theory, at least if we assume, as we have been implicitly, that the gamer is neither psychotic nor being forced by another to play.

In sum, a being is a person according to the Kantian account adopted here if and only if it has the capacities to: gain information about the world; set and pursue ends; strive for coherence among its ends; be self-aware; conform its actions to practical rules, including hypothetical imperatives; and act in accordance with moral imperatives, even when it believes that it would gain more satisfaction by acting contrary to them. If a being possesses all of these capacities, then it can act autonomously according to both the everyday and the split-level notions of such acting.

But under what conditions can a being legitimately be said to possess the capacities constitutive of personhood? According to my view, it suffices for a being to possess them that it has exercised them and that it is possible for it to do so in the future. (By "possible" here I mean practically possible, not merely logically possible.)[60] In principle, a living being from another planet or a non-living artifact such as a sophisticated computer might possess all of the capacities constitutive of personhood. A human being who has died or is alive but whose cerebrum can no longer function is not a person in the sense of the term employed here, on the grounds that he can, practically speaking, no longer exercise the capacities. Let me add that, according to my view, a necessary (but not sufficient) condition for a being possessing the capacities constitutive of personhood at a certain time is that she has exercised at least one of them or is able at that time to exercise the capacity that Kant associates most directly with humanity, namely that to set and pursue ends. It might be the case that, as a result of its natural development, a being will exercise several of the capacities constitutive of personhood and will be able to exercise all of them. But that would not entail that the being now possesses all of these capacities. Beings who have the potential to develop the capacities constitutive of personhood, for example, human zygotes and fetuses, do not thereby count as persons.

The readers of this book are presumably persons. I do not here try to answer the difficult question of when, on average, a normally developing human being becomes a person. But it seems doubtful that newborns are persons.[61] I will also refrain from taking a stand on the question of whether there are non-human animals, for example,

---

[60] Something is logically possible if it is conceivable without contradiction, let us say. It is presumably conceivable without contradiction that a person who has died and whose body has decayed comes back to life and becomes the person he was. So according to the logical sense of possibility, a corpse can exercise the capacities constitutive of personhood. But according to the sense of possibility I wish to employ, namely that of practical possibility, a corpse cannot do this. I will not try here to give a precise definition of practical possibility. But we can say that something is not practically possible if, according to a consensus of scientific experts, it is extremely unlikely to occur.

[61] See Ross Thomson, "The Development of the Person: Social Understanding, Relationships, Conscience, Self," in Nancy Eisenberg, ed. *Handbook of Child Psychology: Social, Emotional, and Personality Development* (Hoboken, N.J.: John Wiley and Sons, 2006).

some chimpanzees or dolphins, who are persons.[62] But it seems clear that cats are not persons.

In the context of Kantian moral theory, the questions of whether infants and animals count as persons seem to be particularly pressing. According to Kant, at least as commonly interpreted, whatever is not a person is a thing: a being that derives any value it has simply by its serving as a means to the ends of persons.[63] If newborns or chimpanzees are not persons, then they are mere things. Akin to tools, they are valuable not in themselves but just insofar as they promote our ends. Many would find unacceptable a theory that had this implication.

But this implication does not follow from the moral principles championed in this book. It develops an account of persons' dignity that leaves open possibilities apparently closed by Kant himself. Kant holds that persons' treatment of non-persons (e.g., non-rational animals) is sometimes morally wrong. But its wrongness is, it seems, solely a function of its impact on persons themselves. Kant claims, for example, that we have a duty not to treat non-rational animals cruelly. But in his view this duty seems to stem simply from the (alleged) fact that by treating such animals in this way, we desensitize ourselves to the suffering of persons, thus making it more difficult for us to fulfill our duties to them and, ultimately, to ourselves.[64] Non-rational animals are mere things, he seems to hold. The account of the dignity of persons developed in the book is consistent with the view that beings other than persons, including but not limited to non-rational animals, have value in themselves, independently of any role they may play in the pursuit of persons' ends. The account is consistent with the view that non-rational animals and human newborns have an unconditional value, beyond mere price.

The account of dignity the book develops does, however, imply that respecting the dignity of persons requires treating them as if they have a value of a higher order than non-persons. So if an agent is to respect the dignity of a person, for example, someone whose life is in imminent danger, she might have to destroy non-persons, if doing so is necessary in order to save the person. But, according to the book, that does not entail that, all things considered, she is morally required to destroy the non-persons. Separating itself from orthodox Kantianism, the book does not hold that we have a categorical obligation to respect the dignity of persons. The book acknowledges that in some cases, failing to respect the dignity of persons is, all things considered, morally permissible.

---

[62] For discussion of the capacities of animals, including evidence that some have self-awareness and even a form of moral agency, see David DeGrazia, *Taking Animals Seriously: Mental Life and Moral Status* (Cambridge: Cambridge University Press, 1996), 166–210.

[63] Wood, *Kant's Ethical Thought*, 143. A basis for this interpretation can be found, for example, at GMS 428.

[64] See MS 443. For an interesting reconstruction of Kant's views on duties with respect to animals, see Lara Denis, "Kant's Conception of Duties Regarding Animals: Reconstruction and Reconsideration," *History of Philosophy Quarterly* 17 (2000).

1.4 Concepts Akin to that of Treating Others Merely as Means

Much of Part I of this book is an attempt to specify the notion of someone's treating another merely as a means. In ordinary discourse, there are, of course, notions that are closely related to that of treating others merely as means—in particular those of manipulating and exploiting them. (I refer, of course, to notions of manipulating and exploiting others according to which doing so is (*pro tanto*) morally wrong.) None of these notions seems univocal; each seems to capture a range of behaviors, with the edges of the range not clearly delineated. As we will find, it is challenging to pinpoint central features of treating others merely as means. The book makes no such attempt with regard to these other two notions. However, it is worth pointing out that although the notions overlap, some intuitively clear cases of exploitation and manipulation are *not* intuitively clear cases of treating someone merely as a means.

Suppose, for example, that a blizzard is forecast for the city, and in order to boost her profits, a hardware store manager doubles the price of the snow-shovels she has in stock. We might say that she is exploiting customers who come in and, no questions asked, buy the shovels at an inflated price. On one influential account, to exploit others is to take unfair advantage of them.[65] And that, arguably, is what the store manager is doing. But I do not believe we would say that she is treating her customers merely as means. Treating another merely as a means, at least in the sense in which doing so amounts (typically) to *acting* wrongly, seems to involve more of an attempt to undermine the other's control over the pursuit of his ends than what the manager does. But, to alter the example, the manager would presumably be treating a customer merely as a means if, in order to assure a sale, she quells suspicions he raises regarding the shovels' high price by telling him falsely that it simply reflects an industry-wide increase in manufacturing costs. The manager's lie aims to diminish the customer's ability to get a lower price elsewhere.

The lying manager's treatment of the customer would, intuitively speaking, count as a case not only of her exploiting him and treating him merely as a means, but also of her manipulating him. The manager reacts to the customer's suspicion that he is about to be overcharged by offering him a reason, which she knows to be false, for relinquishing this suspicion. And that seems to manifest a kind of manipulativeness. According to one sensible account, a manipulator uses bad reasons or arguments, which she presents as good ones, in order to alter another's beliefs or desires.[66] But there are cases of someone's manipulating another that do not seem to amount to her treating the other merely as a means. For example, suppose that someone's friend really wants him to join her in a walkathon to benefit the homeless, but he has plans to attend a concert that day. In an effort to get him to join her, she asks him how he can ignore the

---

[65] Alan Wertheimer, *Exploitation* (Princeton, N.J.: Princeton University Press, 1996).
[66] Claudia Mills, "Politics and Manipulation," *Social Theory and Practice* 21, (1995): 100.

plight of the homeless. "You're better than that," she tells him. And on the basis of their conversation he decides to do the walkathon.[67] Intuitively speaking, she manipulates her friend, but she does not treat him merely as a means.

The upcoming chapter attempts in effect to show that, in light of philosophical shortcomings in a reconstruction of Kant's Formula of Humanity that ignores the Mere Means Principle, the principle warrants further attention. Chapters 3 and 4 specify in detail conditions under which a person does and does not just use another. In the end, these specifications help make up our Chapter 5 account of requirements for respecting the dignity of persons. A core feature of this account (KID) is that the status of persons is such that we ought not to treat them merely as means.

---

[67] Marcia Baron discusses cases like this; see Marcia Baron, "Manipulativeness," *Proceedings and Addresses of the American Philosophical Association* 77 (2003): 43.

PART I

Principles

2

Death, Dignity, and Respect

Philosophers attracted to Kantian ethics have recently followed Kant himself in focusing on the Formula of Humanity (FH) as a basis for specifying what we are morally required to do.[1] This principle commands: "So act that you treat humanity, whether in your own person or in the person of any other, always at the same time as an end, never merely as a means."[2] At bottom, the principle commands us so to act that we always treat humanity as an end in itself.[3]

According to a prominent way of interpreting this principle, namely what I call the "Respect-Expression Approach," we treat humanity as an end in itself just in case our actions express proper respect for the value it possesses.[4] The main claim of this chapter is that if we take the Respect-Expression Approach, FH has problematic normative implications. Sometimes an action is right even if it leads to loss of life. I specify cases in which, many of us believe, withdrawing life-sustaining medical treatment, killing in self-defense, and sacrificing one's life for others are each morally permissible. But on the Respect-Expression Approach Kant's principle yields the conclusion that these actions are wrong. If Kant or contemporary Kantians offered a convincing a priori justification of FH (as interpreted in accord with the Respect-Expression Approach), showing that by making some very plausible general assumption regarding practical rationality or morality we rationally commit ourselves to abide by FH, then we would have sufficient reason to rethink our convictions regarding these cases. But, as I explained in Chapter 1 (1.2), I believe and here assume that no such a priori

---

[1] Allen Wood catalogs the extent to which Kant relies on FH in deriving duties in *Kant's Ethical Thought* (Cambridge: Cambridge University Press, 1999), 139–41. According to J. David Velleman, "Kant was right to say that trading one's person in exchange for benefits, or relief from harms, denigrates the value of personhood, respect for which is a criterion for morality (Kant would say, *the* criterion)," "A Right of Self-Termination?" *Ethics* 109 (1999), 614.

[2] Immanuel Kant, *Groundwork of the Metaphysics of Morals*, trans. Mary Gregor (Cambridge: Cambridge University Press, 1996), 429, italics omitted. I am referring to Preussische Akademie (vol. IV) pagination, which is included in the margins of the Gregor translation. I cite the *Groundwork* as GMS. I have substituted the more familiar "So act that you treat humanity" for Gregor's "So act that you use humanity."

[3] It is widely agreed that an agent's acting so that she treats humanity (in herself or any other) as an end (i.e., end in itself) is a necessary and a sufficient condition for her conforming to FH. See, for example, Thomas Hill, Jr., *Dignity and Practical Reason in Kant's Moral Theory* (Ithaca: Cornell University Press, 1992), 41–2.

[4] See Velleman, "A Right of Self-Termination?" 611, and Wood, *Kant's Ethical Thought*, 141 (discussed below).

justification has been successful. If I am correct that on the Respect-Expression Approach FH has the counterintuitive implications I allege, then we have grounds for skepticism regarding this principle or at least regarding this approach to it. There is room for a new approach to reconstructing FH or elements of it.

The chapter is divided into two main parts. The first (2.1–2.2) examines the content of FH, according to the Respect-Expression Approach. The chapter begins by specifying briefly what Kant likely means when he suggests that humanity is an "end." It then probes the special value he seems to attribute to humanity, namely its dignity. These discussions aim to give plausible (although certainly not definitive) interpretations of Kant. The chapter then explores the Respect-Expression Approach to FH, as developed by Allen Wood. At issue is the content of the Respect-Expression Approach, not the extent to which it squares with Kant's text. The second part of the chapter is an attempt to show, through some rather detailed examples, that the Respect-Expression Approach has problematic practical implications regarding cases of cessation of medical treatment (2.3), but especially regarding cases of self-defense (2.4), and self-sacrifice (2.5).

It is probably already evident that the focus of this chapter is not Kant scholarship. I do not try to determine whether Kant embraced the Respect-Expression Approach to FH or whether FH on this interpretation generates results that harmonize with his ethical views as a whole.[5]

2.1 Humanity as an End and Its Value

Kant employs "humanity" interchangeably with "rational nature."[6] In doing so he suggests that having humanity involves having certain rational capacities. Among them are the capacities to set and pursue ends and to act autonomously, that is, (roughly) to conform to self-given moral imperatives purely out of respect for these imperatives.[7] In what follows, I use the terms "humanity," "rational nature," and "capacity of rational choice" interchangeably.

Kant suggests that humanity is not an "end to be effected," but rather an "independently existing end."[8] In the most general sense, he implies, an end is an object for the sake of which an agent either acts or ought to act.[9] Winning a tennis tournament might count as such an object—it would be an end to be effected—but so might an existing

---

[5] For a strong case against the idea that the Respect-Expression Approach constitutes the best interpretation of Kant, see Oliver Sensen, "Dignity and the Formula of Humanity (ad IV 429, IV 435)," in *Kant's "Groundwork of the Metaphysics of Morals": A Critical Guide*, ed. Jens Timmerman (Cambridge: Cambridge University Press, 2010) and *Kant on Human Dignity* (Berlin: Walter de Gruyter, 2011).

[6] See, for example, GMS 439.

[7] See Hill, *Dignity and Practical Reason*, 38–41.

[8] GMS 437; see also Immanuel Kant, *The Metaphysics of Morals*, trans. Mary Gregor (Cambridge: Cambridge University Press, 1996), 442. I am referring to Preussische Akademie (vol. VI) pagination, which is included in the margins of the Gregor translation. I cite the *Metaphysics of Morals* as MS.

[9] See Wood, *Kant's Ethical Thought*, 116.

object. For example, the United States now has a federal "Respect for flag" law, which states that the US flag "should never touch anything beneath it," "be carried flat or horizontally," "be used for advertising purposes," and so forth.[10] According to those who support this law, the presence of a US flag gives them reasons to do certain things, for example, to try to ensure that it not touch the ground. When they do these things, they presumably do them for the sake of the flag. For them the flag is an end, something for the sake of which they act.[11]

When Kant calls humanity an end in itself, he is not only suggesting that it is something for the sake of which we act or ought to act, but also that it has a particular value.[12] To say that something is an end in itself is to say that it has value with three main features, Kant implies.

First, an end in itself is an objective end, as opposed to a subjective or "relative" end. Objective ends, if there are any, hold for all rational agents. In other words, the idea of securing them makes available to all such agents a ground, that is, a justifying and motivating reason, for acting. But subjective ends do not give all rational beings grounds for securing them.[13] Suppose a particular object is a subjective end. If an agent does not value it, either in itself or as a means to something else, then it has no worth to him. And if the object has no worth to him, intimates Kant, then he does not have a ground to secure it. For him, it is not an end. Kant seems to have the following view: an agent has a ground to secure an object only if he values it or at least is rationally compelled to value it. In the latter case, the agent is presumably able, through rational reflection, to come to value the object, thereby gaining a ground to secure it. From this discussion it should be clear that not all independently existing ends are objective ends. The US flag is an independently existing end, but it is not an objective end. Citizens of China sometimes presumably have no reason to salute in its presence.

Second, an end in itself has absolute or, as we will refer to it, unconditional worth.[14] If something has such worth, then it is valuable under every possible condition, that is, in every possible context, in which it exists. Moreover, the unconditional worth of an end in itself does not diminish no matter what happens to it. A person who loses his zest for life, his fortune, or his sight does not thereby forfeit any of the worth he has as an end in himself. Finally, according to the Kantian notion, if something has unconditional worth, then neither its actions nor their effects diminish this worth.[15]

---

[10] US CODE Title 4, chapter 1, § 8. Respect for flag.
[11] According to US CODE Title 4, chapter 1, § 8 Respect for flag (j), "the flag represents a living country and is itself considered a living thing." When advocates of this law honor the flag, they might, therefore, ultimately be acting both for the sake of the flag (which they consider to have the status of a living thing) and for the sake of the nation the flag represents (a living country).
[12] In FH "end" is equivalent to "end in itself." See GMS 428.
[13] GMS 428, 431.
[14] GMS 428.
[15] Kant says that a good will is good without qualification (GMS 393), which I take to be equivalent to saying that it is unconditionally good. And it is clear that, according to Kant, in all possible circumstances in which it appears, a good will is not only good, but that its level of goodness does not vary according to its

In Wood's words, "the worth of all rational beings is equal."[16] So no matter how many people a criminal harms, he has no less worth than anyone else.

Third, an end in itself has dignity, that is, "unconditional and incomparable worth."[17] We have just noted what it means to have unconditional worth. Kant explains incomparable worth by contrasting it with price: "What has a price can be replaced by something else as its equivalent; what on the other hand is raised above all price and therefore admits of no equivalent has a dignity."[18] The value of something with dignity, then, is incomparable in the sense that it has no equivalent for which it can legitimately be exchanged. That it has no such equivalent seems to have two implications.[19] First, something with dignity can never be legitimately sacrificed for or replaced by something with price. Not even all the gold in Fort Knox would truly compensate for the killing of one rational agent. Second, something with dignity cannot even be legitimately sacrificed for or replaced by something else with dignity.

This position has some striking implications. For example, suppose a tourist piloting a boat can save only the lives of the three strangers stranded on one island or the lives of the five strangers stranded on another. If, according to Kant, it is legitimate for the tourist to save the five, it is not because five persons have greater value than three.[20] Moreover, if in his view it is ever legitimate to kill one being with dignity, thereby saving several other such beings, it is not because it is legitimate to make an exchange of the (lesser) value inherent in the former with the (greater) value inherent in the latter. An end in itself has dignity in that it has unconditional value and nothing, not even a group of other ends in themselves, has greater value.

It is worth emphasizing that, according to Kant, all beings with humanity necessarily also possess dignity. The only way such a being can lose its dignity is by losing its

---

effects. Even if a good will "were completely powerless to carry out its aims; if with even its utmost effort it still accomplished nothing, so that only good will itself remained ... even then it would still, like a jewel, glisten in its own right, as something that had its full worth in itself" (GMS 394).

[16] Wood, *Kant's Ethical Thought*, 132.

[17] GMS 435–6 and MS 434–5, 462.

[18] GMS 434; see also MS 462.

[19] See Hill, *Dignity and Practical Reason*, 47–9.

[20] I think it is legitimate for the tourist to save the five, according to Kant. According to him, beneficence is an imperfect duty. See, for example, MS 452–4. Saving the greater number is, unsurprisingly, compatible with fulfilling this duty. A more challenging question is whether the duty of beneficence, as Kant conceives of it, would *require* the tourist to save the five. I do not think it would, but the grounds for this conclusion are too complex to explore here. In what Frances Kamm calls conflict-free cases, namely ones in which saving some does not preclude saving others, the notion that persons have incomparable value obviously poses no barrier to saving a greater number. Imagine, for example, that eight people are stranded on an island and will die if not rescued. A rescuer can save all eight or save fewer. If the rescuer saves each of the eight on the grounds that, as a person, each one is worthy of saving, she does not in so doing treat anyone's value as comparable to anyone else's. However, the notion that persons have incomparable value does entail that it would be mistaken for a rescuer to save all eight on the grounds that preserving more persons preserves more value in the world. See Kamm's related discussion in Frances Kamm, *Morality, Mortality, Volume 1: Death and Whom to Save from It* (New York: Oxford University Press, 1993), 80–1.

humanity.[21] A person whom others hold in contempt or who even has contempt for himself does not thereby forfeit his dignity.

2.2 The Respect-Expression Approach to FH

In the *Groundwork of the Metaphysics of Morals*, Kant calls rational nature (i.e., humanity) an "object of respect."[22] In the *Metaphysics of Morals*, he suggests that any being with humanity must not only respect himself, but "exacts respect for himself from all other rational beings in the world."[23] It is thus not surprising that Wood tells us the following:

> Though [FH] takes the form of a rule or commandment, what it basically asserts is the existence of a substantive value to be respected. This value does not take the form of a desired object to be brought about, but rather the value of something existing, which is to be respected, esteemed, or honored in our actions.[24]

FH is a moral standard for our actions, that is, for what we intentionally do.[25] According to this principle, an action is morally permissible (in accordance with duty) just in case it expresses proper respect for the worth of humanity, says Wood.[26] As an action-guiding principle, Wood suggests, FH amounts to the following:

> RFH: Act always in a way that expresses respect for the worth of humanity, in one's own person as well as in that of another.[27]

RFH constitutes the core of the Respect-Expression Approach to FH. Of course, RFH is to be understood as a categorical imperative: a principle that all of us (human agents) have an overriding obligation to conform to, regardless of what we might be inclined to do. For the sake of ease of expression, RFH commands that we act always in a way that expresses respect, rather than proper respect for the worth of humanity. But we need to keep in mind that a type of action might express proper respect, or, in short, respect, for the worth of humanity simply by virtue of expressing no disrespect for it. The Respect-Expression Approach does not embrace the idea that every morally permissible type of action involves some positive affirmation of the value of humanity.[28]

In order to derive duties from RFH to act (or refrain from acting) in certain ways we must rely on intermediate premises, according to Wood. For example, he offers the following as the sort of intermediate premise requisite to derive a duty not to make false promises: "Pf: A false promise, because its end cannot be shared by the person to whom

---

[21] Kant's considered view is that dignity is inalienable from humanity, I believe. But there are passages in which he seems to imply that a being can retain its humanity yet forfeit its dignity. For discussion, see Samuel Kerstein, "Treating Oneself Merely as a Means," in *Kant's Ethics of Virtues*, ed. Monika Betzler (Berlin: Walter de Gruyter, 2008), 217–18.

[22] GMS 428.

[23] MS 435; see also MS 462.

[24] Wood, *Kant's Ethical Thought*, 141.

[25] ibid.   [26] ibid. 147.   [27] ibid. 150.

[28] This point stems from personal correspondence with Wood.

the promise is made, frustrates or circumvents that person's rational agency, and thereby shows disrespect for it."[29] The claim in Pf that a false promise shows disrespect for the promisee's rational agency amounts for Wood to the claim that it expresses disrespect for the worth of his humanity. Assuming RFH and Pf are true, it follows that we have a duty not to make false promises. So, in short, according to the Respect-Expression Approach, moral duties to act (or refrain from acting) in certain ways do not stem directly from RFH. This principle must be coupled with intermediate premises: ones that specify whether some sort of conduct expresses respect for the worth of humanity. (If intermediate premises are necessary to derive from RFH conclusions regarding the moral permissibility of types of actions, then they are obviously also necessary to derive such conclusions regarding particular actions.)

Several points regarding Wood's characterization of intermediate premises warrant attention. First, they are "logically independent" of RFH in the sense that the truth of this principle does not itself guarantee the truth of any such premise.[30] That we ought to act always in a way that expresses respect for the worth of humanity does not itself entail that any particular sort of conduct in fact expresses or fails to express such respect. Second, according to Wood the intermediate premises are "hermeneutical": "they involve interpreting the meaning of actions regarding their respect or disrespect of the dignity of rational nature."[31] For example, Pf above incorporates an interpretation of the action of making a false promise to someone, namely that it expresses disrespect for the worth of this person's humanity.

That these intermediate premises are hermeneutical does not entail that there are no standards that a legitimate one must meet, Wood underscores. Our interpretations of what actions express regarding the worth of humanity are subject to rational argument.[32] So we can presumably show that some such interpretations are to be rejected. Some mischaracterize what an action expresses. For example, an injured person asks an emergency room physician to stop his bleeding. The doctor does so in a minimally painful and invasive way, with no motive other than a desire to restore his health. It is plainly incorrect to say that the doctor's action expresses disrespect for the value of the patient's humanity, as RFH construes this value. Other intermediate premises are unacceptable on the grounds that they falsely imply that a being possesses humanity, for example, "Killing a Japanese maple expresses disrespect for the value of its rational agency."

Wood suggests a further constraint on the legitimacy of intermediate premises, one that emerges from reflection on an interpretation of RFH he rejects. One might think that an agent's treatment of another expresses respect for his humanity just in case the treatment is accompanied by a respectful state of mind. In other words, what the agent does conforms to duty if and only if when doing it he has the thoughts or feelings it is appropriate to have when treating an end in itself in some way. Wood rejects this view.

---

[29] ibid. 153.   [30] ibid. 152.   [31] ibid. 154.   [32] ibid. 154–5.

According to him, some actions, such as making a promise one does not intend to keep, express disrespect for the value of humanity no matter what the mindset of the agent who performs them. Moreover, some actions express respect for its value even if, in performing them, the agent is devoid of a respectful state of mind.[33] When you buy a cup of coffee from someone, you might have no respectful thoughts or feelings toward her. Your mind might be (almost) entirely elsewhere. But that does not entail that you have failed to treat her as an end in herself. For your action might nevertheless express respect for the worth of the person's humanity, Wood suggests.

But do some actions express respect for the value of humanity even if an agent does them with a disrespectful state of mind? Let us suppose that the person buying coffee is a gangster. His tone of voice and body language are neither noticeably polite nor impolite. But not only is he devoid of respectful thoughts or feelings toward the seller, he feels contempt for her and reflects that he would not hesitate for a moment to kill her if that would further his purposes.[34] Would it be legitimate to conclude that the buyer's action expressed disrespect for the worth of the seller's humanity solely on the basis of his thoughts and feelings? Wood suggests that it would not be: "in dealing honestly with you, I treat you with respect whatever my inner state may be," he says.[35] If the gangster's action amounts to his paying the seller the posted price in order to get a cup of coffee, then I think it would, in Wood's view, express respect. For despite his contemptuous feelings and reflections, the gangster has "dealt honestly" with the seller. In contrast, suppose that a kind-hearted customer who feels respect for the seller and would under no circumstances do her bodily harm steals coffee. His action would nevertheless fail to express respect for the worth of the seller's humanity. So Wood suggests the following constraint on legitimate intermediate premises: An intermediate premise's conclusion that conduct expresses respect or disrespect for the value of humanity must not be grounded simply on an agent's having or not having a certain mindset in carrying it out.

In any case, according to the Respect-Expression Approach to FH, this principle asserts the existence of a value to be respected. Consider an action that Kant judged to be wrong, say a person's committing suicide in order to avoid a painful illness. What makes the action wrong is that it fails to respect the value of humanity. An action fails to respect this value just in case it fails to express respect for it. But any action that fails to express respect for the value does so at least in part by suggesting an inaccurate message regarding what this value is. A full account of what makes this suicide wrong would necessarily include the notion that it expresses a false message, namely that some person (i.e., the one about to kill himself) does not have dignity. Moreover, if an action expresses such a message, then it expresses disrespect for the value of humanity and is morally impermissible.

---

[33] ibid. 117.
[34] Derek Parfit suggests this example in *On What Matters*, Vol. 1 (Oxford: Oxford University Press, 2011), 216.
[35] Wood, *Kant's Ethical Thought*, 117.

Before turning to criticism of RFH, let us take note of a claim Wood has recently made regarding morally appropriate responses to the value of humanity. Our being bound to act in a way that expresses respect for the value of humanity does not entail that we are bound to *preserve* humanity, he insists:

> If it is normally a requirement of morality that we should seek to preserve rational beings in their existence, then this is a *consequence* of the fact that if an existent being has basic and unconditional value, then the state of affairs of its continued existence also has great value, at least most of the time. But from the fact that humanity or rational nature has dignity, or fundamental and unconditional value, it by no means follows that the value of human *life* is basic or unconditional.[36]

In one sense it should be uncontroversial that humanity's having dignity fails to entail that human life has dignity. For some human life (e.g., the life of a severely brain-damaged and permanently unconscious accident victim) is devoid of rational nature and thus devoid of dignity in the Kantian sense at issue. But Wood is trying to make a different point here. According to him, there are circumstances in which respecting the value of the rational nature in a person requires that the person herself or another end her life and thus extinguish her rational nature. Wood does not specify in any detail what those circumstances are. His remarks remain very general, as in the following: "At times people are in terrible situations where living up to the dignity of their rational nature even requires them to sacrifice their continued existence."[37] In such situations, a person's "sacrificing" her existence would presumably express respect for the value of her humanity, according to Wood.

In this chapter, I am neither assuming nor trying to establish the impossibility of such situations. More generally, I am neither assuming nor trying to show there to be no circumstances in which respecting the value of the rational nature in a person requires that the person herself or another kill her and destroy that rational nature.

Here is what I am doing. First, I am supposing that if a person's life is extinguished, then so is his rational nature. Second, I am taking as a *defeasible presumption* that destroying rational nature fails to express respect for its dignity. Actions that end the existence of the humanity in a person tend to send messages incompatible with the notion that his humanity has unconditional and incomparable worth. (Think of cases such as intentionally killing an innocent person solely for the sake of monetary gain or of avoiding time in prison.) There may well be exceptions to this tendency, but their status as such requires explanation.[38] Third, I am claiming that in certain cases in which

---

[36] Allen Wood, *Kantian Ethics* (Cambridge: Cambridge University Press, 2008), 86–7.

[37] ibid. 87.

[38] I mention a potential exception in note 45. Two other potential exceptions are suggested by cases Kant discusses. In illustrating his notion of "pure morality" Kant describes a man who refuses to calumny an innocent person, even though, as he is aware, the price of this refusal is his own death. See Immanuel Kant, *The Critique of Practical Reason*, trans. Mary Gregor (Cambridge: Cambridge University Press, 1996), 155–6. I am referring to Preussische Akademie (vol. V) pagination, which is included in the margins of the Gregor translation. I cite *The Critique of Practical Reason* as KpV. It might be said that this man's action

many of us are convinced that withdrawing life-sustaining medical treatment, killing in self-defense, and sacrificing one's life for others are morally permissible, RFH implies the contrary. In other words, I am claiming that *in these cases*, which I describe in detail below, RFH has counterintuitive implications.

I discuss many objections to this claim, several of which attempt to explain how the life-ending action in question actually expresses respect for the value of humanity. I contend that the objections fail. It is not always easy to discern whether an action expresses such respect. It can be difficult to determine whether it sends a message that is compatible with the view that humanity has dignity—not because the Kantian notion of dignity is vague, but rather because what an action "means" is subject to interpretation. The message that an action (or at least an action under a certain description) expresses can presumably vary with historical and cultural context. In one context, someone's spreading dirt on his face might express the message that he is undeserving of the love of God (who is believed to be in heaven), but in a different context it might suggest the notion that he is deserving of close contact with God (who is believed to be one with the earth).[39] It is important to acknowledge the possibility of contextual variation in the message an action expresses. But this variation does not, I believe, come into play regarding my interpretations of the actions I discuss below. These interpretations assume that the actions are performed in contemporary contexts familiar, at least via journalistic accounts, to readers of this book. Of course, some philosophers might reject the interpretations. But these philosophers should either explain why their interpretations are more plausible than the ones I offer or embrace the skeptical view that no rational argument concerning the meaning of actions is possible. According to the skeptical view, the Respect-Expression Approach renders FH too indeterminate to be usable. I do not embrace that view. I contend not that the Respect-Expression

---

constitutes a sacrifice of his rational nature, but that it nevertheless expresses respect for the value of this nature. His refusal expresses such respect, it might be said, because it is required by his rational nature's highest principle, namely the moral law. Second, in a discussion of punishment, Kant describes a "man of honor" convicted of having taken part in an attempt to overthrow a government: someone like Baron Balmerino who participated in the Scottish rebellion. Kant asks us to suppose that the court enables the rebel to choose as his punishment either death or convict labor. Kant affirms that he would choose death: for he "is acquainted with something that he values even more highly than life, namely honor" (MS 334). In choosing death, it might be claimed, the rebel would be sacrificing his rational nature, but also expressing respect for its dignity. The truth of this claim, as well as of the analogous claim regarding the calumny case, is debatable, I believe. To cite just one issue, the rebel's choosing death might express respect for his dignity in some sense of the term. For example, it might express respect for his dignity in the sense of his high social standing: a status (supposedly) far above that of anyone who would engage in manual labor. But why should we conclude that his choice would express respect for his dignity defined as the unconditional and incomparable value of his rational nature? Does the supreme principle of that nature really require him to choose death?

[39] To point to another example, putting on blackface might send a very different message depending on whether a Caucasian or an African-American is doing so.

Approach makes FH too vague to be functional, but rather that on this approach FH has acutely counterintuitive normative implications.[40]

2.3 Withdrawal of Medical Treatment

If the "meaning" of an action is inconsistent with the view that humanity has dignity, then the action expresses disrespect for the value of humanity and thus conflicts with RFH. ("Act always in a way that expresses respect for the worth of humanity, in one's own person as well as in that of another.") Many of us believe that certain actions (specified below) are morally permissible. However, RFH yields the conclusion that they are not, for the actions in question convey a message that humanity does not have unconditional and incomparable value, or so I argue.[41]

A patient has ALS (Lou Gehrig's disease), a lethal malady the progressive symptoms of which include muscle weakness, paralysis, and eventual loss of the ability to speak, swallow, and breathe. He is on a respirator, and has no hope of living off of it. He might survive for as little as a few months or as long as a couple of years. As is typical among people with (even advanced) ALS, the patient's mind is sharp. Moreover, he does not suffer from clinical depression. As psychiatrists have confirmed, he easily meets standards of mental competence. His senses of sight, touch, and hearing are unaffected, and technology enables him to communicate and manipulate his environment. He sets and effectively pursues various ends. But he finds his condition intolerable. The patient's doctor, who knows the details of his mental and physical condition, is the only person who as a practical matter can turn off the respirator. Over several months, the patient persistently asks her to do this so that his suffering will come to an end.

Suppose that the doctor turns off his respirator. She intends the patient to die in order that he no longer suffer. Philosophers differ on whether the doctor counts as killing the patient, as opposed to letting him die.[42] But we need not enter this debate here. In any case, according to many of us the doctor's action is morally permissible. But according to the Respect-Expression Approach, crystallized in RFH above, FH implies that it is not.

RFH commands that we always treat people in a way that expresses respect for the worth of their humanity. It should be obvious that the patient does indeed possess humanity. Since the onset of his disease, the range of ends he can set and reasonably

---

[40] Of course, if, contrary to my opinion, the Respect-Expression Approach does make FH too indeterminate to be usable, then there is room for a different approach to FH (or parts of it). And that is a result I welcome.

[41] In presenting the cases, I assume that the reader and the persons described in the cases believe that if a human being who has humanity dies, then that being's humanity goes out of existence. In short, I assume that they do not believe that rational nature persists in an afterlife.

[42] For example, Dan Brock ("Voluntary Active Euthanasia," *Hastings Center Report* 22 (1992), 12–13) implies that in this case the doctor would count as killing the patient, while Frances Kamm ("A Right to Choose Death," *Boston Review* Summer (1997), section V) suggests that she would count as letting him die, at least if it was she (or staff at her hospital) who put the patient on the respirator in the first place.

expect to realize has significantly diminished. But he maintains his reason and his ability to act on self-given principles.

Moreover, the patient's humanity has the same unconditional and incomparable value it had before he got sick. Although, as a result of his suffering, the patient himself believes that it is in his interest to die, the value of his humanity diminishes not at all. J. David Velleman, who seems to embrace a view that amounts to something like the Respect-Expression Approach to FH, writes: "The dignity of a person is a value that differs in kind from his interest." The patient might hold that his life is no longer worth living. But, Velleman continues, a person's "dignity is a value on which his opinion carries no more weight than anyone else's."[43] Even the patient's request to die has no effect on the value of his humanity. In at least some circumstances, including I think those of our example, such a request suggests the patient's denial that his humanity has dignity. But such a denial does not entail any actual loss of value. Echoing Velleman once again, a person's dignity is not a value for him, but rather a value in him.[44]

Whether it is correct to say that the doctor kills the patient or that she lets him die, she terminates his aid, intending that he die and thereby suffer no more. Her action does not express a message consistent with the view that the patient's humanity is unconditionally and incomparably valuable. (Here again is an intermediate premise in Wood's sense.) Instead it sends the message that in the context of this patient's suffering and his insistence that she do something that will result in its cessation, his humanity is simply not worth preserving. The action also suggests that the patient's humanity is less valuable than it was in an earlier context, namely one in which his suffering was much less intense and he did not want to die. For in the earlier context, the doctor did not intend the patient's death. So RFH implies that it is morally wrong for the doctor to turn the respirator off. To many of us, this implication is counterintuitive.

Let me now consider replies to this objection, each of which aims to show that since the doctor's action does not express disrespect for the value of humanity, RFH does not imply that it is wrong.

First, an opponent might reply as follows: Humanity is the capacity of rational choice. Actions express respect for the value of this capacity if they express respect for its exercise. The patient exercises the capacity of rational choice in setting the end of being free from suffering and pursuing it by trying to bring about his own death. The doctor's turning off the respirator sends a clear message that she respects the patient's exercise of his capacity of rational choice. So, the reply concludes, her action also sends the message that his capacity of rational choice is valuable.

This reply invokes an unacceptable premise, namely that if actions express respect for the exercise of the capacity of rational choice, then they express respect for the value of the capacity itself. An example will help us to see that this premise is unacceptable.

---

[43] Velleman, "A Right of Self-Termination?" 611. For evidence that Velleman accepts something like the Respect-Expression Approach to FH, see 624.

[44] ibid. 613.

Suppose that an aging poet reasonably believes the following: His poetry, which explores themes of violence and redemption, is deep and important, but underappreciated. The most efficient and perhaps the only way to draw significant attention to his work, which is by far the most important thing to him in his life, is for him to die a violent death. The poet is unable to shoot himself, so he asks an acquaintance to do it. He offers the acquaintance impeccable evidence that he is mentally competent and persistently renews his request. Now suppose that the acquaintance shoots and kills the poet. His action expresses respect for the poet's exercise of his capacity of rational choice, namely his pursuing the end of getting his work noticed through his dying a violent death. But it fails to send the message that this capacity itself is worth preserving. For the acquaintance's action wipes this capacity out of existence. Suppose that instead of acquiescing to the poet's request, the acquaintance refuses to shoot him and repeatedly tries to convince him that bringing attention to his work is not worth the loss of his life. This action does not express respect for this particular exercise of the capacity of rational choice by the poet—it thwarts it. But the action does express respect for his capacity itself. The capacity of rational choice (humanity) is something over and above any particular exercise of this capacity. An action's expressing respect for the latter is not a sufficient condition (nor is it a necessary one) for its expressing respect for the value of the former.

One might embrace this conclusion, yet insist nevertheless that the doctor's turning off the ALS patient's respirator does express respect for the value of his humanity. For in this particular case, one might claim, an action's expressing respect for the exercise of the capacity of rational choice does suffice for its expressing respect for the value of the capacity itself.

But what is the justification for this claim? Granted, in some cases one's expressing respect for a particular exercise of the capacity of rational choice might amount to expressing respect for the value of the capacity itself. Suppose a person lends a colleague money based on his promise to repay her by a specific date. (As the borrower is aware, the lender needs him to repay her on time so that she can make the down-payment on a home.) And in fact the borrower does repay the lender on time. Let us assume that the borrower's action of paying back the loan on time expresses respect for the lender's exercise of her capacity of rational choice, that is, for the exercise of agency involved in making the loan. Perhaps, then, it thereby also expresses respect for the value of the lender's capacity of rational choice, that is, her humanity, itself. Someone might hold the doctor's action to be analogous to the borrower's. But I fail to see good reason to take this position. Both the borrower's action and the doctor's, let us grant, express respect for some particular use another puts to her own capacity of rational choice. But unlike the borrower's action, the doctor's will, as she is fully aware, result in the destruction of another's capacity of rational choice. That is in short why the doctor's action expresses a message contrary to the idea that the patient's rational nature is unconditionally and incomparably valuable.

Some might be attracted to the view that an action's expressing respect for the patient's exercise of his capacity of rational choice amounts to expressing respect for the value of this capacity itself for the following reason. They might assume that his capacity is, as it were, exhausted by this particular exercise of it. They might envisage the patient as so obsessed with trying to put an end to his own suffering as to be incapable of any other exercise of the capacity of rational choice. But that is simply not the patient as I have described him. He can and does pursue other ends. His capacity of rational choice has not collapsed into one use of it. So we are left with the conclusion that even though the doctor's action expresses respect for a particular exercise by the patient of his capacity of rational choice, her action expresses disrespect for the value of the capacity itself.[45]

A second reply according to which the doctor's action does not express disrespect for the value of the patient's humanity takes shape against the background of some remarks by Velleman:

> When a person cannot sustain both life and dignity, his death may indeed be morally justified. One is sometimes permitted, even obligated, to destroy objects of dignity if they would otherwise deteriorate in ways that would offend against that value. The moral obligation to bury or burn a corpse, for example, is an obligation not to let it become an affront to what it once was. Librarians have similar practices for destroying tattered books— and honor guards, for destroying tattered flags—out of respect for the dignity inherent in these objects . . . These examples suggest that dignity can require not only the preservation of what possesses it but also the destruction of what is losing it, if the loss would be irretrievable . . . Respect for an object of dignity can sometimes require its destruction.[46]

The notion of dignity Velleman is using when he discusses objects such as flags is wider than the strict Kantian notion we have been employing, according to which an object has dignity just in case it has unconditional and incomparable value. Such objects have dignity, Velleman suggests, just in case they have value that is to be respected, as opposed to maximized, or in Kant's language just in case they have value as "independently existing" ends rather than as ends "to be effected."[47] Velleman is not, of course, claiming that it is always wrong to trade off the value inherent in an American flag or a Bible for that inherent in something else, say a person. I will discuss Velleman's remarks as they relate to objects such as flags and as they relate to persons and might constitute a reply to my analysis of the case of the ALS patient. I am not sure what Velleman would say about this case. So although the reply (developed below) to my analysis of it is

---

[45] If a patient were capable of exercising his capacity of rational choice only in connection with his attempt to end his life, then perhaps a doctor's action of turning off his (life-sustaining) respirator would express respect for the value of his capacity of rational choice. Perhaps we would have a case in which respecting the value of the rational nature in a person requires that the person herself or another end her life and thus extinguish her rational nature. For discussion, see 5.5 below.

[46] Velleman suggests this in "A Right of Self-Termination," 617 and 626.

[47] ibid. 617.

inspired by his comments, I do not claim that it captures his own view. In any case, the reply is not promising, or so I argue.

Velleman's discussion suggests the following reasoning concerning "objects of dignity" such as flags. When a national flag becomes irreparably tattered it typically loses its capacity to serve as a fitting emblem of the nation and thus loses at least some of the value it had as such an emblem.[48] Burning a flag that is *about to* irretrievably lose some of its value can express respect for it, goes the argument. It can send a message that as a result of its possessing value as such an emblem, not just any demise will do. Allowing it to forever lose its value through a process of further decay might "offend against" what it is and the value it possesses.

We can formulate roughly parallel reasoning regarding an ALS patient. As a result of having this disease, a patient can lose his humanity (i.e., rational nature) and thus some of his value. Acting with the intention of bringing about the death of an ALS patient who is about to lose his humanity can express respect for the value of his humanity. It can convey that as a result of his possessing humanity, and thus unconditional and incomparable value, not just any demise will do. Allowing such a patient to forever lose his value through the disease process might "offend against" his humanity and the value it possesses. In our example, the doctor's acting with the intention of bringing about the death of the particular patient described, one who is competent and who has asked to die, *does* express respect for the value of his humanity. For the doctor's not doing so and thereby allowing the patient to forever lose his value by succumbing to ALS would offend against his dignity as a person, concludes this argument.

This argument constitutes a reply to the contention that RFH condemns the doctor's action in our original example. Although the patient has not lost his humanity he is about to. Precisely how long before he loses it no one knows: it could be a month, it could be years.

In any case, the reply is not convincing. While undergoing the process of physical degradation that will lead to his losing his humanity, the patient might feel embarrassed, ashamed, and humiliated. His sense of self-worth might diminish. When this happens, a patient is sometimes said to have experienced a loss of dignity.[49] But a *sense of* self-worth is not dignity, as we have been employing the notion. A patient has dignity in our usage regardless of how high or low he believes his worth to be. He has dignity, that is unconditional and incomparable value, just as long as he has humanity. If the doctor's refraining from pulling the plug and allowing the patient's disease to destroy his humanity would "offend against" or constitute an "affront to" the patient's humanity, it would *not* do so by actually diminishing its value. It would have to do so by sending the message that his humanity failed to have unconditional or incomparable value.

---

[48] In some cases a tattered flag might be more effective than a pristine one at serving as an emblem for a nation, for example, if the tattered flag remained flying during a successful defense against foreign attack.

[49] See, for example, H. Chochinov et al., "Dignity in the Terminally Ill: a Cross-Sectional, Cohort Study," *Lancet* 360 (2002), 2029.

How would it send that message? Granted, in refraining from acting on the patient's request, a doctor might not be expressing respect for a particular exercise of the capacity of rational choice. But, as we have noted, failure to express respect for an exercise of this capacity need not amount to a failure to express respect for the value of the capacity itself. And it does not seem to in this case where for practical purposes expressing respect for the exercise of the capacity involves acting with the intention that the capacity cease to exist.

Indeed, it is acting with this intention that sends the message that humanity is not unconditionally and incomparably valuable. ALS is going to rob the patient of his humanity, but it has not done so yet. The doctor's turning off the respirator and intentionally killing the patient or, if one prefers, intentionally letting him die, suggests the idea that in the context of this patient's suffering and his insistence that she do something that will result in its cessation, his humanity is not worth preserving.[50]

2.4 Self-Defense

It is a commonplace to hold that killing another in self-defense is, in some circumstances, morally permissible.[51] Consider the following case: A law-abiding journalist has discovered widespread financial improprieties in a large company. He has a well-grounded suspicion that a security officer employed by this company aims to kill him in order to keep him from revealing what he knows. The officer follows him into an enclosed alley and approaches him with knife raised. The journalist tries, to no avail, to reason with him. He then takes out a gun and yells at the officer to stop. But he continues to move forward, now just a step away. The journalist reflects in a flash that the officer is a former paramilitary soldier and an expert in hand-to-hand combat. He concludes, very reasonably, that if he doesn't shoot to kill, he is very unlikely to escape from the situation alive.

---

[50] A roughly parallel point applies in connection with the flag. According to my construal of Velleman's remarks, burning a flag that is *about to* irretrievably lose some of its value can express respect for it. But this position does not seem to cohere with the US Code on "Respect for Flag." For this states the following: "The flag, *when it is in such condition that it is no longer a fitting emblem for display,* should be destroyed in a dignified way, preferably by burning" (US CODE Title 4, chapter 1, § 8. Respect for flag. (k), emphasis mine). If a flag is now a fitting emblem for display, then burning it would seem to express disrespect for its value—even if it will soon no longer be a fitting emblem. In 1989, before the US Supreme Court in effect ruled their enforcement unconstitutional, 47 states as well as the federal government had "flag protection" statutes. See John Luckey, "Flag Protection: A Brief History and Summary of Recent Supreme Court Decisions and Proposed Constitutional Amendment," in *Congressional Research Service Report for Congress* (2005).

[51] Hill remarks ("Respect for Humanity," *The Tanner Lectures on Human Values* (1997), <http://www.tannerlectures.utah.edu/lectures/documents/Hill97.pdf>, 72–3) that "lethal force in self-defense is permitted by traditional moral standards". According to Frances Kamm (*Morality, Mortality, Volume II: Rights, Duties, and Status* (New York: Oxford University Press, 1996), 129), "it is permissible to kill a malicious aggressor in self-defense."

If the journalist intentionally kills the officer in this case, he acts in self-defense and his action is morally permissible, or so many of us believe.[52] But in shooting and killing the officer, the journalist would not express respect for the value of his humanity. (This claim is an intermediate premise in Wood's sense). Since humanity's value is unconditional, what a person does cannot diminish it; since humanity's value is incomparable, it cannot be outweighed by the value of anything else. The journalist's action of destroying the officer's humanity fails to express respect for the very special value it possesses.

In order to see this, consider first a case of intentionally killing (and thus destroying the humanity of) an innocent, non-threatening person who wishes to remain alive. Doing so would express disrespect for the value of her humanity. It would convey a message that her humanity falls short of having unconditional and incomparable value. But now note that the value of the security officer's humanity neither disappears nor at all diminishes when he acts as a malicious aggressor. That it maintains its full value is just part of what it means to say that it has unconditional value in the Kantian sense. In destroying the officer's humanity, the journalist is destroying something with no less worth than the humanity of an innocent, non-threatening person. If someone's intentionally killing the innocent person expresses disrespect for the value of her humanity, then the journalist's intentionally killing the officer expresses disrespect for the value of his. So, according to the Respect-Expression Approach, the journalist's intentionally killing the officer in self-defense is morally impermissible.[53]

Now let me consider several objections to this argument. First, one might acknowledge that on the Respect-Expression Approach an agent's having a certain mindset in acting constitutes neither a necessary nor a sufficient condition for his action's expressing disrespect for humanity. One might insist, however, that an agent's mindset can be crucial to correctly assessing what his action expresses. If the journalist has a certain mindset in killing the officer, then his action does not express disrespect for the value of the officer's humanity, one might claim. It does not express disrespect if, for example,

[52] This belief is not unusual among philosophers writing on issues surrounding morality and self-defense. David Rodin (*War and Self-Defense* (Oxford: Oxford University Press, 2002), chapters 1–4) implies that he would endorse it as does Suzanne Uniacke (*Permissible Killing: the Self-Defence Justification of Homicide* (Cambridge: Cambridge University Press, 1994), chapter 5). As both Rodin and Uniacke suggest, some philosophers believe that if the journalist intentionally kills, as opposed to unintentionally but foreseeably kills, the officer, then the journalist acts wrongly. Like Rodin and Uniacke, I believe that it can sometimes be permissible to intentionally kill in defense of one's life.
[53] One might wonder whether, if, as I claim, the Respect-Expression Approach to FH implies that the journalist acts wrongly, it would also imply that some cases of a civil authority's carrying out the death penalty on a convicted (and guilty) murderer would be wrong. I believe that this approach to FH would indeed also imply this, even in some cases in which the execution was "humane," although I do not attempt to defend this view here. If I am right, then, in light of Kant's championing of capital punishment (MS 331–337), we would have some, but far from decisive, evidence against the Respect-Expression Approach to FH as a historically accurate representation of Kant's views. In any case, that on this approach FH implied that some cases of "humane" capital punishment of convicted murderers was morally impermissible would not in my view count against the plausibility of the principle itself.

he is committed to killing the officer only on condition that doing so is in all likelihood necessary to save his own life and he feels deep sadness when this condition is realized.

The journalist's having this mindset might suggest something about his character, for example, that it is better than that of someone who in the same situation would dispatch the officer without feeling a thing. But consideration of the journalist's mindset fails to warrant the conclusion that his action expresses no disrespect for the value of the officer's humanity. No matter what his emotions might be and no matter what he has committed himself to doing if circumstances were different, it remains that, against the attacker's will, he intentionally destroys his humanity. This action conveys the message that the officer's humanity falls short of having unconditional and incomparable value.[54]

A second objection begins with the claim that the right of a person not to be killed is, under normal circumstances, a consequence of his humanity's special value as invoked in RFH. But a person can forfeit this right. When he has forfeited it, doing what would otherwise constitute a violation of this right does not express disrespect for the value of his humanity. When he tries to murder the journalist, the officer forfeits his right not to be killed. So the journalist's intentionally killing him in self-defense neither violates his right not to be killed nor expresses disrespect for the value of his humanity, concludes the objection.

In response, some prominent accounts of the legitimacy of self-defense do maintain that a would-be assassin such as the officer loses his right not to be killed. But we are focusing specifically on the implications of RFH (or, more precisely, RFH coupled with plausible intermediate premises). The officer maintains his humanity, and the value of it remains undiminished, even as he tries to commit murder. To deny the latter is to deny that humanity has unconditional value. According to the objection, the right of a person not to be killed is, under normal circumstances, a consequence of the value of his humanity. But if the value of the officer's humanity has diminished not at all, then the ground of his right not to be killed remains unaltered: it is fully present before as well as after he attempts to murder the journalist. So the objection's claim that in making this attempt he forfeits his right not to be killed is baseless.

Of course, someone might maintain that the officer's right not to be killed stems from some principle other than RFH, and that when he attempts murder he does indeed forfeit that right. The journalist's intentionally killing him in self-defense might not run afoul of this other principle. But it would run afoul of RFH. The journalist

---

[54] A further objection to the conclusion that the journalist fails to express proper respect for the worth of the officer is the following: "According to Kant, since a murderer has willed his own death, capital punishment shows respect for the will (capacity of rational choice) of the murderer. Similarly, in trying to kill the innocent journalist, the officer has willed his own death. So in intentionally killing him, the journalist shows respect for officer's capacity of rational choice." In response, the objection rests on a (common) misinterpretation of Kant. He explicitly *rejects* the view that a murderer wills his own death (see MS 335). Moreover, even if Kant had embraced this view, why should we follow him?

wipes the officer's humanity out of existence and thereby expresses the message that it falls short of being unconditionally and incomparably valuable.

A third objection to my conclusion that the Respect-Expression Approach has counterintuitive implications also contends that the view stops short of implying that it is morally impermissible for the journalist to intentionally kill the officer. The objection grants that, if the journalist does so, his action expresses disrespect for the officer's humanity. But it makes a further claim, namely that, according to the Respect-Expression Approach, if the journalist does *not* try to kill the officer, then his action expresses disrespect for the worth of his own humanity. For if he does something less than what will maximize his chances for survival, his action conveys that his humanity does not have dignity, the objection continues. Whether the journalist tries to kill the officer or not, he expresses disrespect for someone's humanity. In a case in which whatever one does, one expresses disrespect for someone's humanity, it is morally permissible to act in whichever way one chooses, according to the objection. So, contrary to my conclusion, it would be morally permissible for the journalist to kill the officer in self-defense.

The objection depends on the claim that, according to the Respect-Expression Approach, if the journalist fails to attempt to kill the officer, then his action expresses the message that his own humanity does not have the value of an end in itself. But this claim is questionable. That the journalist refrains from trying to kill the officer does not entail that he does nothing to save his own life. Although he reasonably believes that unless he shoots to kill the officer, he will very likely die by his knife, he realizes that this outcome is not certain. He can take steps, short of shooting to kill, in order to survive. He might, for example, try to disable (but not kill) the officer by shooting him in the leg and then struggle with all his might to escape. The journalist's making such an attempt is certainly not necessary in order for him to count as exercising morally permissible self-defense, many of us believe. But this action would, it seems, express respect, not only for the worth of his own humanity, but also for the worth of the officer's.[55] So might a less violent action, such as a renewed effort to reason with the officer and to convince him that it would be wrong, or at least imprudent, for him to stab him to death. These actions would suggest that both the journalist's own humanity and the officer's are worth preserving. Granted, if the journalist intentionally does nothing in order to save himself, as opposed, say, to doing nothing as a result of being frozen with fear, and if his doing nothing counts as an action, then his action might express disrespect for the value of his humanity. But, as we have seen, the journalist's

---

[55] A philosopher raising the third objection might claim that the journalist's intentionally shooting the officer in the leg would also express disrespect for the value of the officer's humanity. But, if true, this claim would render RFH even less plausible than I have been charging. For RFH would fail, not only to affirm the moral permissibility of intentionally killing an attacker in defense of one's life, but also fail to affirm the moral permissibility of intentionally disabling (but not killing) an attacker in defense of it.

refraining from trying to kill the officer is compatible with his making some attempt to save his own life.

Even if, contrary to my contention, whatever the journalist did short of shooting to kill would express disrespect for the value of his own humanity, the objection falls short. For it rests on the claim that if, whatever one does, one expresses disrespect for one person's humanity, it is morally permissible to act in any way one chooses. But this claim clashes with the Respect-Expression Approach itself. RFH commands: "Act always in a way that expresses respect for the worth of humanity, in one's own person as well as that of another." If an action expresses disrespect for the worth of humanity, then it is morally wrong. That is the case regardless of whether, whatever one does, his action will fail to respect the worth of humanity.

An opponent who raises this third objection embraces the idea that the journalist faces a moral dilemma. Since shooting to kill is morally impermissible, he is morally required not to shoot; since refraining from shooting to kill is also morally impermissible, he is also required to shoot. Whatever he does, he will violate a moral requirement. (Of course, Kant himself rejected the possibility of there being any genuine conflict of duties.[56] So, *if* Kant accepted the Respect-Expression Approach to FH, he should not accept this third objection to the conclusion that the journalist would be wrong to shoot to kill in self-defense.)

At this point, someone might grant it to be inconsistent with RFH to claim that in cases in which whatever one does one expresses disrespect for someone's humanity, whatever one does is morally permissible. Modifying the third objection, she might insist instead that in such cases RFH, charitably interpreted, fails to yield a verdict on the moral status of the actions one can perform. In other words, it has what Thomas Hill, Jr., calls "gaps": "it provides no way, even in principle, to determine what one should, or even may, do" in these cases.[57] She would, of course, need to offer some justification for concluding that in these cases RFH leaves us with a gap rather than with a dilemma. She might appeal to the principle, embraced by Kant, that if we morally ought to do something, then we can do it (ought implies can).[58] According to RFH, our actions ought always to express respect for the value of humanity. But in the cases in question, our actions cannot do this, for they will fail to express respect for the value of at least one person's humanity, she might contend. If RFH applied in such cases, it would violate the principle that ought implies can. Therefore, RFH does not apply in these cases and thus yields no verdict on what should be done in them. It has gaps. So the principle does not imply that the journalist's killing the officer is morally impermissible.

---

[56] For helpful discussion of his rejection, see Hill, *Human Welfare and Moral Worth: Kantian Perspectives* (Oxford: Clarendon Press, 2002), 370–4.

[57] Hill, *Human Welfare and Moral Worth*, 379. Hill does believe that FH leaves some gaps. But it is not clear whether he thinks it does so in the journalist/officer scenario.

[58] See, for example, KpV 125 and 159.

This modified objection does not succeed.[59] For it does not threaten the position defended above, namely that the Respect-Expression Approach implies that some actions available to the journalist would not express disrespect for anyone's humanity and so would be morally permissible. An example of such an action would be the journalist's trying to disable, but not kill, the officer in order to escape from him, even though he knows he will not likely succeed.[60] If this is correct, then there is no gap in RFH such that it fails to apply to the journalist in this scenario. Moreover, even if there were such a gap, RFH would fail to yield a conclusion that many of us embrace, namely that it is not wrong for the journalist to intentionally kill the officer.[61]

2.5 Heroic Self-Sacrifice

Our last case requires less discussion than the previous ones. Below is a narrative accompanying the award of a military honor, the Silver Star, to a US army private killed in combat in Iraq:

> PFC Ross McGinnis' platoon was conducting a combat patrol to deny the enemy freedom of movement in Adhamiyah [Northeast Baghdad] and reduce the high-level of sectarian violence in the form of kidnappings, weapons smuggling, and murders... PFC McGinnis was manning the [machine gun] on the Platoon Sergeant's [vehicle]. His primary responsibility was to protect the rear of the combat patrol from enemy attacks. Moments after PFC McGinnis' vehicle made [a] turn traveling southwest a fragmentation grenade was thrown

---

[59] The attempt to justify the view that in the journalist case RFH yields no moral prescription is problematic in the context of Kant interpretation. If RFH has gaps then, contrary to what Kant claims, it is not a viable candidate for the supreme principle of morality. For part of what he means by calling the principle supreme is that it be able to determine regarding every action whether or not it is morally permissible. I defend this last claim in Samuel Kerstein, *Kant's Search for the Supreme Principle of Morality* (Cambridge: Cambridge University Press, 2002), 17–18.

[60] Take a different case, namely one in which, as the journalist knows with apodictic certainty, there is no way he could save his own life without killing the officer. The journalist refrains from killing the officer and so gets killed. Does the journalist's action fail to respect the value of his own humanity? Perhaps it does, even if, had there been something he could have done that might have saved his own life without killing the officer, the journalist would have done it. If his action does fail to respect the value of his own humanity and if, as I charge, his killing the officer would fail to express respect for the value of the officer's, then RFH would either yield a moral dilemma or suffer from a gap. Either way, it would not help to generate the verdict that many of us would affirm, namely that the journalist's killing the officer would in this case be morally permissible.

[61] An additional objection one might raise against my conclusion here is the following: "The journalist's intentionally killing the officer expresses respect for the value of the officer's humanity since his action accords with principles to which all rational beings are committed." But why should we believe that the journalist's action accords with principles to which all rational beings are committed? It would presumably accord with such principles if it accorded with RFH; for RFH is supposed to be the supreme principle of morality. But it would obviously be circular simply to assume that his action accords with RFH; for that is precisely what is at issue. Second, why should we assume that if the journalist's action does accord with principles to which all rational beings are committed that it thereby expresses respect for the unconditional and incomparable value of the officer's rational nature? Perhaps all rational beings are committed to the principle of maximizing their own welfare. But surely an action that accords with that principle, for example, one of killing another to avoid one's own financial ruin, need not express respect for the dignity of the other's rational nature.

at [it] by an unidentified insurgent from an adjacent rooftop. He immediately yelled "grenade" on the vehicle's intercom system to alert the four other members of his crew. PFC McGinnis made an attempt to personally deflect the grenade, but was unable to prevent it from falling through the gunner's hatch. His Platoon Sergeant, the truck commander, was unaware that the grenade physically entered the vehicle and shouted "where?" to PFC McGinnis. When an average man would have leapt out of the gunner's cupola to safety, PFC McGinnis decided to stay with his crew. Unhesitatingly and with complete disregard for his own life he announced "the grenade is in the truck" and threw his back over the grenade to pin it between his body and the truck's radio mount. When the grenade detonated, PFC McGinnis absorbed all lethal fragments and the concussion with his own body killing him instantly. His early warning allowed all four members of his crew to position their bodies in a protective posture to prepare for the grenade's blast. As a result of his quick reflexes and heroic measures, no other members of the vehicle crew were seriously wounded in the attack. His gallant action and total disregard for his personal well-being directly saved four men from certain serious injury or death. PFC McGinnis' extraordinary heroism and selflessness at the cost of his own life, above and beyond the call of duty, are in the keeping of the highest traditions of military service. He gallantly gave his life in the service of his country.[62]

Many of us believe that PFC McGinnis' action was, at the very least, morally permissible. But RFH implies the contrary. According to the narrative, PFC McGinnis' action of throwing his back over the grenade showed "total disregard for his personal well-being" and "complete disregard for his own life." The narrative implies that PFC McGinnis intentionally allowed himself to be killed in order to save his comrades, or at least that in order to save them, he did something a virtually certain consequence of which was his own death. Either way, his action expressed disrespect for the incomparable worth of his humanity. (Here we have an intermediate premise.) It sent the message that its value was not as great as the value of that of the four other soldiers taken together.[63] So, according to RFH, PFC McGinnis' action was wrong.[64]

---

[62] "Ross Andrew McGinnis: Specialist, United States Army," Arlington National Cemetery, <http://www.arlingtoncemetery.net/ramcginnis.htm>. In addition to the Silver Star, PFC McGinnis was awarded a Medal of Honor, the United States' highest military distinction. See "Medal of Honor Recipients: Iraq," US Army Center of Military History, <http://www.history.army.mil/html/moh/iraq.html.U.S>.

[63] Here one might object that the message of PFC McGinnis's action might not include the idea that the value of his humanity is not as great as that of the four other soldiers. For he might have decided to throw himself on the grenade not on the basis of reasoning that the value inherent in four persons is greater than that inherent in one, but rather, say, on some (very quick) random procedure that determined which unconditionally and incomparably valuable being(s) he would try to preserve. Given this possibility, we cannot conclude that the message of his action includes the notion that persons have comparable worth, ends the objection. In light of the quoted narrative, I find this interpretation of the meaning of PFC McGinnis's action to be very implausible. But even if it were on target, it would still be the case that, although he had no duty as a soldier to do so, he intentionally destroyed his own humanity or at least did something a virtually certain consequence of which was its destruction. And either way, his action conveyed the message that his humanity lacked dignity.

[64] In the "casuistical questions" Kant raises after he discusses the duty not to commit suicide, he poses the following question: "Is it murdering oneself to hurl oneself to certain death (like Curtius) in order to save one's country?—or is deliberate martyrdom, sacrificing oneself for the good of all humanity, also to be

Here one might object that rather than implying that his action was wrong, the Respect-Expression Approach implies nothing regarding it. In other words, PFC McGinnis' action falls through a gap in RFH. But it is implausible to conclude that RFH provides no way, even in principle, to determine what is morally permissible in this case. According to RFH, it would seem permissible for PFC McGinnis to do what the narrative suggests "an average man" would do, namely after trying to deflect the grenade and to warn his comrades about it, to "leap out of the gunner's cupola to safety." These actions would express no disrespect for the value of PFC McGinnis's own humanity. In trying to deflect the grenade and thereby carrying out his duty as a soldier, he would be accepting some risk to his life. But he would obviously neither be intentionally sacrificing it, nor doing something a virtually certain consequence of which would be his own death. Moreover, the actions would express no disrespect for the value of his comrades' humanity. First, no more than in the case of what he actually did would PFC McGinnis himself be threatening the lives of his comrades; the person who tossed the grenade was doing that, of course. Second, the action of leaping to safety would take place against the background of an already-made attempt by PFC McGinnis to save the others. It would be inaccurate to characterize his behavior as a whole as abandonment. PFC McGinnis's good-faith effort to prevent the grenade's killing his comrades would convey a message consistent with the notion that their humanity had unconditional and incomparable value. Contrary to the objection, appeal to RFH would, I believe, imply something determinate about actions open to the private, namely that it would be morally permissible for him, after trying to save his fellow soldiers, to leap to safety.

Even if the objection succeeded in establishing that RFH really implied nothing about the case of PFC McGinnis, that would not absolve it of the criticism at issue. For, if I am correct, according to reflective common sense a plausible principle would enable us to conclude that what the private did was morally permissible, as opposed to impermissible (wrong).

A further objection one might raise to my application of RFH to this case is the following: Suppose that PFC McGinnis believed that he was morally required to jump on the grenade. One lives up to the dignity of one's rational nature if and only if one does what one believes to be morally required, according to the objection. So PFC McGinnis' jumping on the grenade expresses respect for the value of his rational nature. In short, PFC McGinnis was in one of those terrible situations Wood refers

considered an act of heroism?" (MS 423–4). If we take the Respect-Expression Approach to FH, then in my view such a self-sacrificial act does turn out to be "murdering oneself" and thus to be morally impermissible. In the "Notes on the Lectures of Mr. Kant on the Metaphysics of Morals" taken by Johann Friedrich Vigilantius, we read: "It is permissible to venture one's life against the danger of losing it; yet it can never be allowable for me deliberately to yield up my life, or to kill myself in fulfillment of a duty to others; for example, when Curtius plunges into the chasm, in order to preserve the Roman people he is acting contrary to duty . . . " Immanuel Kant, *Lectures on Ethics*, trans. Peter Heath (Cambridge: Cambridge University Press, 1997), 629. I am referring to the Preussische Akademie edition (vol. XXVII) pagination, which is included in the margins of the Heath translation.

to (2.2), that is, one in which a person's honoring the dignity of his rational nature requires him to sacrifice its continued existence.

This objection suffers from serious shortcomings. First and most obviously, it rests on the assumption that PFC McGinnis believed that he had a moral duty to jump on the grenade. But what if, as seems plausible, he did not believe this? (He might have taken jumping on the grenade and thereby saving his colleagues to be an excellent thing to do, but not something that it would be wrong to refrain from doing.) Then the objection does not allow us to avoid the implausible conclusion that, according to RFH, his jumping on the grenade was morally impermissible.

Second, it is simply false that one honors the dignity of one's rational nature (as the Respect-Expression Approach conceives of this dignity) just in case one does what one *believes to be* morally required. What if someone takes it to be morally required to kill himself in order to avoid the humiliation of losing his ability to walk? That act, even if he does it just because he believes it to be his duty, would not express respect for the unconditional and incomparable value of his rational nature. In response to this point, the objector might claim that one lives up to the dignity of one's rational nature just in case one actually does what is morally required. The argument would then be that since PFC McGinnis's jumping on the grenade was morally required, his action of destroying his rational nature expressed respect for its value. But according to many of us, PFC McGinnis's action was not morally required. Indeed, it seems to be a paradigmatic example of supererogatory conduct. It was above and beyond the call of duty—both duty in the sense of what is required of a soldier and duty in the sense of what is required of a person in general.

The cases we have examined are ones in which actions that lead to someone's death seem to many of us to be right (morally permissible). But appeal to RFH requires us to conclude that the actions, as well as any similar to them which also fail to express respect for the value of humanity, are wrong, or so I have tried to show.[65]

Success in showing this, even assuming as I have throughout that no viable a priori justification of RFH is at hand, does not completely discredit the Respect-Expression Approach to FH. Someone might reject the view that any of the actions described are morally permissible. For such a person this chapter has simply worked out some implications of the Respect-Expression Approach. But many of us hold that it is not wrong for the doctor to turn off the ALS patient's respirator, perhaps still more that it is not wrong for the journalist to intentionally kill the company security officer; and nearly everyone, I believe it is safe to say, thinks that what PFC McGinnis did was morally permissible. (I suspect that our confidence in these views tends to increase progressively from the first to the last; for example, we are more secure in our belief in the moral permissibility of PFC McGinnis's action than we are in that of the doctor's.)

---

[65] In Chapter 6, I illustrate that the Respect-Expression Approach also has implausible implications in some cases in which we must decide to whom to give a scarce, life-saving resource.

As I discuss in Chapter 5, one might modify the Respect-Expression Approach in an effort to avoid counterintuitive implications. One might, for example, envisage humanity to possess unconditional but not incomparable value. But such modifications do not yield promising results, I argue. Rather than trying to bolster the Respect-Expression Approach to FH, I reconstruct the constraint FH contains against treating others merely as means (Chapter 3). I then build an account of the dignity of persons according to which they have a special status, one such that they ought not to be treated merely as means and such that they ought to be treated as having unconditional, transcendent value (Chapter 5). This account aims to capture much of what many of us find attractive in Kantian notions of dignity, without having to assume all of the problematic implications of the Respect-Expression Approach. The present chapter has left open the possibility that there is no reconstruction of FH or elements of it that has more plausible normative implications than does the Respect-Expression Approach. But much of the remainder of this book is dedicated to showing that this is not the case.

One final note: although it is implausible to claim, as does the Respect-Expression Approach, that an action is wrong if it expresses disrespect for humanity conceived of as unconditionally and incomparably valuable, this approach suggests an important point, one that we revisit in Chapter 7. We do indeed seem to have a reason (albeit not a decisive one) to refrain from doing things that express the idea that persons have the value of mere tools.

3

The Mere Means Principle

According to the Mere Means Principle, it is wrong to treat others merely as means. If the argument of Chapter 2 is convincing, then a leading approach to the Formula of Humanity (FH), an approach that de-emphasizes this principle, suffers from serious shortcomings. It thus makes sense to try other approaches to FH, or at least to elements of it, including an approach that attempts to develop the prescription not to treat others merely as means into a clear and plausible principle.

Of course, Kant does *not* hold that if a person refrains from treating another merely as a means, then her action is morally permissible. A person can, for example, act wrongly in Kant's view by expressing contempt for another, even if she is not using him at all.[1] She would be acting wrongly by failing to treat the other as an end in himself, rather than by treating him merely as a means. A person might also act wrongly by treating *herself* merely as a means.[2] Refraining from treating others merely as means is a necessary, but not a sufficient condition for acting rightly, according to Kant. So in Kant's own view, a specification of the Mere Means Principle could never capture all of FH. Following Kant, I embrace the idea that moral requirements are not exhausted by the Mere Means Principle. Although this book does not attempt a complete reconstruction of FH, it does (in Chapter 5) develop ethical requirements inspired by elements of FH in addition to the Mere Means Principle.

The present chapter explores various attempts to set out a *sufficient condition* for treating others merely as means. These attempts are all informed, but not limited, by what Kant actually says in the *Groundwork of the Metaphysics of Morals*. At least three distinct ways of formulating such a condition might, with some plausibility, be gleaned from the *Groundwork*. One might hold that an agent treats another merely as a means if the other cannot share the end she is pursuing in using him (3.2). Or one might contend that she treats him merely as a means if he is unable to consent to her using him. One might interpret this inability to consent in two different ways, namely in terms of it being irrational to consent (3.3) or, rather, in terms of the lack of an

---

[1] See Immanuel Kant, *The Metaphysics of Morals*, trans. Mary Gregor (Cambridge: Cambridge University Press, 1996), 462–4. I am referring to Preussische Akademie (vol. VI) pagination, which is included in the margins of the Gregor translation. I cite the *Metaphysics of Morals* as MS.

[2] See, for example, MS 430.

opportunity to consent (3.4). Using materials suggested by the *Groundwork*, this chapter tries to develop a plausible sufficient condition for an agent's treating another merely as a means, namely the Hybrid Account (3.5). Finally, the chapter considers some objections to this account (3.6).

Before turning to various formulations of sufficient conditions for treating others merely as means, some preliminary points are in order. In this chapter, we are trying to elaborate a Kantian idea of treating others merely as means, central to which is the notion that to treat others in this way is to *act* wrongly. A proposed sufficient condition for using others merely as means will be plausible only if it pinpoints actions that are both recognizable by reflective common sense as such and that are (at least typically) wrong.

According to a "rough definition" suggested by Derek Parfit we use another merely as a means if we both use the other and regard him "as a mere instrument or tool: someone whose well-being and moral claims we ignore, and whom we would treat in whatever ways would best achieve our aims."[3] So, for example, a kidnapper treats his victim merely as a means if she uses him to get money and considers him simply as a tool that she would treat in any way necessary in order to get it.

As Parfit contends, if *this* is how we understand treating others merely as means, then the Mere Means Principle seems highly implausible. According to the principle, in using others merely as means, we are *acting* wrongly, that is, doing something morally impermissible. But we can easily imagine cases in which, on this account, someone would be treating another merely as a means, but would, intuitively speaking, not be acting wrongly. Parfit suggests the case of a radical egoist, someone who would do anything to anyone if he judged that it would best promote his interests.[4] The egoist risks his life to save a child from drowning with the sole end of getting a significant reward. In doing so, he is, according to Parfit's account, treating the child merely as a means. But we reject the conclusion that in saving the child he acts wrongly. We judge that what he does is morally permissible, even though his attitude towards people is morally repugnant. Since Parfit's account of treating others merely as means implies that the Mere Means Principle is false, it makes sense to try to develop a different account of doing so, one that both to a significant extent coheres with ordinary thinking and that pinpoints *actions* that, according to reflective common sense, are wrong.

Parfit, it seems, claims to capture *the* ordinary sense of when someone treats another merely as a means. This sense, he appears to believe, is encapsulated in the "rough definition" quoted above, as well as in examples, discussed below, regarding which we would allegedly conclude that an agent is *not* treating another merely as a means (3.6). Other accounts, he implies, fail to capture the ordinary sense, but instead arrive at a

---

[3] Derek Parfit, *On What Matters*, Vol. 1 (Oxford: Oxford University Press, 2011), 213 and 227.
[4] Parfit, *On What Matters*, 216.

"special," quasi-technical sense of using others merely as means.[5] In my view, there is no single ordinary way in which we think of treating others merely as means. Parfit might capture one set of ideas associated with reflective common sense, but there is another set according to which it *is* plausible to hold that to treat someone merely as a means is to *act* (*pro tanto*) wrongly, or so the chapter tries to show.

According to Kant, the Mere Means Principle allows of no exception, and it thus likely fails to square with our considered moral views. Perhaps no matter how we specify the notion of treating others merely as means, we are able to imagine an extreme scenario in which treating them in this way will not seem to us to be wrong. For example, what if millions of people will die in a nuclear explosion unless we kill one innocent person as a means to preventing it, thereby treating him merely as a means? We resist the conclusion that we would be acting wrongly.

Those of us attracted to the Mere Means Principle should, I believe, revise it by making it non-absolute. We should hold that although we always have significant moral reason not to treat others merely as means, this reason can be overridden. Treating others merely as means is always wrong *pro tanto*, but not necessarily wrong, all things considered. Such a revision should allow us to maintain something like the Mere Means Principle even in the face of cases such as the one just mentioned. Below (3 as well as 5.6) I will make some suggestions regarding when the constraint implicit in the Mere Means Principle gets overridden. In any case, from this point on we will understand the Mere Means Principle to state that it is wrong *pro tanto* for an agent to use another merely as a means.

But if we understand the Mere Means Principle to identify *pro tanto* wrongs, do we not rescue it from Parfit's criticism? In treating another merely as a means, *as Parfit understands this notion*, are we not *acting pro tanto* wrongly? No, we are not. Returning to an example invoked earlier (2.2), a gangster who considers a barista a mere tool to get coffee and who would treat her in whatever way would best serve his interests does not, all things considered, *act* wrongly in simply buying coffee from her. If his buying coffee was wrong *pro tanto*, then there would have to have been some reason he had to buy it that outweighed his moral reason not to buy it. But what was that reason? That he needed a caffeine lift would surely not have sufficed. In order to render the Mere Means Principle plausible, we need to develop a notion of what treating another merely as a means amounts to that differs from the notion Parfit proffers.

A final preliminary point: accounts of treating others merely as means should minimize appeal to concepts of morality or rationality that are just as elusive as our notion(s) of treating others merely as means. If they rely on such concepts, they are unlikely to be illuminating.

---

[5] ibid. 227.

3.1 Using Another

Before examining accounts of agents treating others *merely* as means (i.e., "just using" them), we need to specify when agents count as treating others as means (i.e., using them) at all. Such treatment is often morally permissible. Students treat professors as means to knowledge; professors use students to test their theories, and so forth. Of course, in saying "John used me," a person might mean that John treated her *merely* as a means and thus wrongly. But in my terminology, saying that someone used another or, equivalently, treated the other as a means does not itself imply any negative moral evaluation of his action.

When do we count as using or treating another as a means? This question, which is preliminary to that of when we count as treating another *merely* as a means, poses more challenges than might immediately be apparent. Let me begin, however, by making some relatively uncontroversial observations. It does not seem sufficient for us to count as using another as a means that we benefit from what the other has done.[6] If, on her usual route through the park, a jogger gets pleasure from the singing of a stranger who happens to be passing by, she does not appear to be treating the stranger as a means. Second, not all cases of an agent's intentionally doing something in response to another are cases of his using the other as a means. If someone smiles at a friend who is approaching, for example, he might not thereby be using the friend at all: he might simply be expressing affection.[7]

There are various ways to understand what is to count as using another as a means. Frances Kamm asks us to imagine that "we must choose whether to save A or B, where B but not A will serve the useful function of saving C if he survives."[8] Suppose that we choose to save B partly on the grounds that he will save C if he survives. According to Kamm, we then treat A as a means. We do so "because we consider whether he could be of use to us and reject him because he is not." So let us say that, according to the Kammian understanding, an agent treats another as a means if she chooses whether or not to do something to the other based on her perception of the other's usefulness in bringing about her (the agent's) ends.

If a defender of the Kammian understanding assumes, as we have been, that using another as a means and treating another as a means amount to the very same thing, then this understanding coheres little with ordinary usage (no pun intended). Our talk of using or treating a person as a means surely has affinities with our discourse regarding using or treating things as such. Suppose a homeowner goes through a tool box,

---

[6] See Robert Nozick, *Anarchy, State, and Utopia* (New York: Basic Books, 1974), 31–2.

[7] In what follows, I take the expressions "treat another as a means" and "use another as a means" to be equivalent. For the sake of brevity, I frequently shorten the expression "use another as a means" to "use another." But in order to avoid confusion, I do not shorten "treat another as a means" to "treat another." For it is possible to treat another in some way, for example, to smile at him as an expression of affection, without treating him *as a means*.

[8] Frances Kamm, *Morality, Mortality, Volume 1: Death and Whom to Save from It* (New York: Oxford University Press, 1993), 111.

searching for something with which to drive in a nail. He sees a screwdriver, and quickly focuses his attention elsewhere. He then locates a hammer, which he takes from the box. It would be very odd indeed to say that the homeowner uses the screwdriver as a means, even though he does consider whether it could be of use to him and rejects it because it cannot be. Or consider a basketball player, namely a guard, watching the end of a game from the bench and hoping to get into the action. The coach decides to put in a forward instead, since he judges that the situation calls for an excellent rebounder. It would be a mistake to say that the coach used the guard as a means, even though, on the basis of his judgment that he would not be a suitable means to his end of winning the game, he left him on the bench. In fact, after the game, especially if his team lost, the player might complain to a teammate: "The coach should have used me." Using someone as a means or treating him as such (in the sense in which the two are equivalent) involves more than taking an instrumental attitude toward him or even, based on such an attitude, refraining from taking an action regarding him. It involves (positively) doing something with, or at least something to, some aspect of the person.

We might then say that all cases of using another, or, equivalently, treating another as a means are ones in which an agent intentionally does something to someone in order to secure, or as a part of securing, one of his ends. For example, I use a taxi driver if I hail his cab in order to get to the cinema; I treat my spouse as a means if I lie to her so that her birthday party will be a surprise; and I treat a mugger as a means if I punch him in order to escape from his grasp.[9] A soldier who sets off a bomb solely in order to kill enemy combatants might foresee that innocent bystanders will be harmed. But if he does not intentionally do anything to the bystanders, then he does not treat them as means.[10]

But like the Kammian account, this account seems to count too much as treating another as a means. Suppose that a police officer in pursuit of a suspect pushes a bystander out of his way. The account we are considering implies that the officer has used the bystander as a means, for he has intentionally done something to her (i.e., pushed her aside) in order to attain an end (i.e., to apprehend the suspect). But that implication seems implausible. The officer has treated the bystander in some way, namely as an obstacle to be removed. But he has not used her. To cite an analogy, suppose someone is reaching into a toolbox for a hammer. If a screwdriver is in her way and she shunts it aside to get to the hammer, she has done something to the screwdriver, but she hasn't used it.

In order for an agent to count as using another, it is not enough that she do something to the other in order to realize some end of hers. She must also intend the presence or participation of some aspect of the other to contribute to the end's

---

[9] I employed this account in Samuel Kerstein, "Treating Others Merely as Means," *Utilitas* 21 (2009), 166.

[10] Of course, that the soldier does not treat the bystanders as means does not entail that his setting off the bomb is morally permissible.

realization.[11] The police officer does not intend the bystander's presence or participation to play any role in his capturing the suspect. He thinks of her simply as "in the way." In sum, an agent uses another (or, equivalently, uses or treats another as a means) if and only if she intentionally does something to or with (some aspect of) the other in order to realize her end, and she intends the presence or participation of (some aspect of) the other to contribute to the end's realization.

For purposes of this account, let us understand an agent's intentionally doing something to another in a broad sense so that it can include refraining from performing an action. According to this understanding, an agent can intentionally do something to another by, say, refusing to shake his hand. And it is not hard to imagine cases in which doing this to someone amounts to treating him as a means. Suppose, for example, that during a televised ceremony a citizen refuses to shake a politician's hand in order to convey to spectators that he strongly disapproves of a policy the politician advocates. The citizen intends the politician's presence to contribute to his attaining his end. It seems reasonable to say that the citizen is treating the politician as a means.

An agent uses another through using her capacities. For example, I get directions from a stranger in order to find the train station. In this case, I use his aptitude to engage in means–end practical reasoning. In more Kantian terms, I use his ability to construct a hypothetical imperative—a principle such as "if you want to go from here to the train station, you ought to take your first left, then your second right." Another way for an agent to use another as a means is for her to use his emotional capacities. Political leaders, for example, sometimes try to invoke fear in their constituents (e.g., fear of possible terrorist attacks) as a step toward realizing their aim of changing the law (e.g., gaining legal permission for warrantless wiretapping of their citizens' conversations). There are, of course, also cases in which an agent uses another's physical capacities. When a trafficker in human organs extracts a kidney from a kidnapping victim, he uses the victim as a means. With no pretence to exhaustiveness, we have distinguished three ways an agent might use another, namely through using the other's rational, emotional, or physical capacities.[12] We have, of course, left important questions unanswered regarding when an agent uses another. We might wonder, for example, whether she uses him if she uses information derived from his body (e.g., the results of medical tests) or a part of his body (e.g., a biological specimen). We discuss such questions in Chapter 8, which focuses on research ethics.

Derek Parfit might reject this account of treating another as a means. He asks us to suppose that a doctor is trying to determine whether a patient has a broken rib. She presses in various places on his chest and asks the patient to tell her when it hurts.

---

[11] This account is based on, but not identical with, one developed by Thomas Scanlon in *Moral Dimensions: Permissibility, Meaning, Blame* (Cambridge, Mass.: Belknap Press of Harvard University Press, 2008), 106–7.

[12] One might, for example, use another through using his sense of humor. But it might turn out that a sense of humor is a unique capacity, that is, not reducible to a combination of emotional, rational, or physical capacities.

According to Parfit, the doctor is not treating the patient as a means. But the account I have sketched obviously implies that she is doing so. Parfit suggests that an agent does not use another when the other consents to what she does to him and her aim is to benefit him.[13]

In response, it does sound a bit odd to say that the doctor is using the patient. But this oddness falls far short of giving us good reason to abandon the account. First, the oddness might be explained by a tendency in ordinary language to speak in terms of "using another" when we mean treating him *merely* as a means. When we refer to a doctor's "using" a patient, we sometimes have in mind a negative moral judgment of the doctor's action—a judgment that, at least without further reflection, seems incompatible with the patient's having consented to what the doctor does and her doing it for the patient's sake. Second, Parfit's suggestion regarding conditions under which an agent does not count as using another has shortcomings of its own. It does not sound strange to say that a doctor uses a patient in a research study, even if it is true both that the patient consents to be in the study and that the doctor has enrolled him ultimately because she thinks he might benefit greatly. We can easily imagine the doctor saying: "I've got some good news; I can use you in the study." I suspect that no account of using another is going to square entirely with ordinary usage.

Parfit might suggest an additional reason for rejecting our account. He asks us to consider a case in which Brown attacks Green with a knife, trying to kill him, but Green saves himself by kicking Brown in a way that predictably breaks his leg.[14] Parfit then says that "just as we do not *use* falling rain when we wear raincoats to protect ourselves from being drenched, we do not *use* the people who attack us when we protect ourselves from their attack."[15] Does our account imply that Green uses his attacker, Brown? If, as seems plausible, we describe Green's end as something like "not being harmed," then it does not seem that the account would imply this. While Green does intentionally do something to Brown, namely break his leg, he does not intend Brown's presence or participation to contribute to his end of not being harmed. As far as Green's attaining that end goes, it would be better if Brown were not there at all. But in other cases, our account would imply that a person defending himself against an attack was using his attacker. Suppose that two muggers approach you, and you push one of them into the other in order to gain an opportunity to escape. On our account, you would be treating someone as a means; for you would intentionally be doing something to someone (e.g., pushing him) in order to attain some end (e.g., securing an opportunity to escape), and you would intend the person's presence to contribute to your attaining the end. This implication strikes me as plausible: in pushing the mugger, you are treating him as a means. Indeed, the case seems to me to cast doubt on the truth of the general claim that we do not use the people who attack us when we protect ourselves from their attack.

[13] Parfit, *On What Matters,* 213 and 222.   [14] ibid. 221.   [15] ibid. 221–2.

3.2 End Sharing

Much discussion of what it means to treat others merely as means stems from a single passage in the *Groundwork*. Kant is attempting to demonstrate that FH generates a duty not to make false promises:

> he who has it in mind to make a false promise to others sees at once that he wants to make use of another human being *merely as a means*, without the other at the same time containing in himself the end. For, he whom I want to use for my purposes by such a promise cannot possibly agree to my way of behaving toward him, and so himself contain the end of this action.[16]

We are looking for an interpretation of this passage that will help us to understand in general terms what it means to treat another merely as a means. According to Allen Wood, Kant is arguing here that making a false promise to another would treat the other merely as a means since it would express disrespect for the worth of his rational nature. "A false promise, *because its end cannot be shared by the person to whom the promise is made*, frustrates or circumvents that person's rational agency, and thereby shows disrespect for it."[17] Apparently, according to Wood, when Kant says that a promisee cannot "himself contain the end" of a false promisor's action, he is intimating that the latter cannot share the promisor's end. That interpretation seems reasonable enough. Borrowing from Wood here, we might try to construct a sufficient condition for using others merely as means. Although Wood does not do so himself, we might claim that if another cannot share an agent's end in using her in some way, then the agent treats the other merely as a means.

Two agents presumably share a particular end if the following is the case: they are both trying, or at least have both chosen to try, to realize this end. If this is not the case, then they presumably do not share the end. But what, precisely, does it mean to say that two agents *cannot* share an end? Returning to the example at hand, what does it mean to say that the promisee cannot share the promisor's end? Wood is not helpful on this question. From the outset it is important to specify precisely which of the promisor's ends the promisee cannot share. It is presumably the promisor's end of getting money from the promisee without ever paying it back. For the promisor's *ultimate* end might be that of diminishing world hunger, and there seems to be no reason why the two cannot share that end. But it remains unclear just what sense of "cannot" Kant is invoking (or should invoke) in suggesting that a promisee cannot share a false promisor's end.

On one reading, Kant holds that the promisee cannot share the promisor's end in that it is logically impossible for him to do so.[18] Suppose the promisor, a borrower, has

---

[16] Immanuel Kant, *Groundwork of the Metaphysics of Morals*, trans. Mary Gregor (Cambridge: Cambridge University Press, 1996), 429–30. I am referring to Preussische Akademie (vol. IV) pagination, which is included in the margins of the Gregor translation. I cite the *Groundwork* as GMS.

[17] Allen Wood, *Kant's Ethical Thought* (Cambridge: Cambridge University Press, 1999), 153, italics added.

[18] Thomas E. Hill, Jr., *Human Welfare and Moral Worth* (Oxford: Clarendon Press, 2002), 69–70.

the end of getting money from the promisee, a lender, without ever paying it back. The borrower makes a false promise in order to secure that end. At the time he makes a loan on the basis of this promise, the lender cannot himself share the end of the borrower's getting the money from him without ever paying it back, goes this reading. If the lender shared the borrower's end, then he would not really be making a loan. For according to our practice, it belongs to the very concept of making a loan, as opposed, say, to giving money away, that one believe that what one disburses will be repaid.

Given our aim of arriving at a plausible account of treating others merely as means, this interpretation of the promisee's inability to share the promisor's end is unhelpful. First, it just does not seem to be logically impossible for the lender to share the borrower's end. The borrower, let's say, is trying to secure the end of his getting money from the lender without ever repaying it in order ultimately to enjoy a vacation in Tahiti. The lender is also trying to realize this end—not so that the borrower can enjoy a vacation in Tahiti, but so that he, the lender, who despises the borrower, can revel in the demise of the borrower's reputation. The joy the lender would experience at the borrower's loss of reputation would more than compensate for his loss of the money, the lender might think. The lender shares the borrower's end. As far as I can tell, nothing in this case entails that in making the loan, the lender fails to believe that what he gives out will be repaid. It is easy to imagine him reflecting, with regret, that the borrower *will* probably pay him back and that his ultimate aim of enjoying the demise of the borrower's reputation will probably remain unfulfilled. It might not be a common occurrence, but it is logically possible for a lender to share a borrower's end of getting money from the lender without ever paying it back.

Second, this interpretation of the false promising case leads naturally to the view that a sufficient condition for an agent's treating another merely as a means is that it is logically impossible for the other to share the end the agent is pursuing in treating her in some way. However, this sufficient condition would be anemic: it would fail to register as treating others merely as means not only the case of false promising just discussed, but other paradigmatic cases of doing so as well. Take, for example, a loiterer who threatens an innocent passerby with a gun in order to get $100. The sort of sufficient condition for treating another merely as a means that we seek should allow us to conclude that the loiterer is treating the passerby merely as a means; for he is mugging her. But the sufficient condition on the table does not do this. It is improbable, but still logically possible, that the passerby shares the loiterer's end of his getting $100.

One might reply that if the loiterer is to count as mugging the passerby and thereby treating her merely as a means, it must not be the case that the passerby shares the loiterer's end. For if she does, then he simply does not count as mugging her. It follows from the concept of mugging that the victim does not have the same end that the mugger is pursuing.

But this reply fails. Suppose the passerby is the loiterer's nephew whom she hasn't seen in a few years and that on this dark, foggy night the two do not recognize one another. The aunt (the passerby) has the end of her nephew (the loiterer) having $100, which is coincidentally precisely the amount of cash she has in her purse. She was planning to give the $100 to him for his birthday the next day. In these circumstances, it is surely possible for him to mug her in order to realize his end of getting $100. How else would we describe his waving a gun in her face and shouting to her to give him her purse? It is logically possible for two agents to share an end even in cases paradigmatic of one's using the other merely as a means. So the logical possibility account of end sharing would yield a weak sufficient condition for an agent's treating another merely as a means.[19]

According to a different interpretation of Kant, another cannot share the end an agent pursues in using him in some way if how the agent behaves "prevents [the other]

---

[19] We can imagine a defender of the logical possibility criterion insisting that a mischaracterization of the agents' ends in the examples above is what is responsible for the criterion's appearing to be implausible. The loiterer's end, claims this defender, must be described as something like "obtaining $100 through getting this passerby to fear that if she does not give it up, she will be harmed." If his end is described in this way, then, the defender maintains, it is indeed logically impossible for the passerby to share it. According to him, someone cannot share another's end if it is logically impossible for her to intentionally contribute to the end's realization. The defender insists that it is logically impossible for the passerby to intentionally contribute to the realization of the loiterer's end; for her pursuing it would make it the case that if the loiterer obtained the money, it would not be through getting her to fear that if she does not give it up, she will be harmed. The defender of the logical contradiction criterion would also insist that the false promisor's end be described differently, for example, as "the promisee's handing over money on the grounds of his mistakenly trusting an untrustworthy repayment assurance." It is logically impossible for the promisee to share this end, the defender claims. For if he pursued it, he would in effect make it the case that if he hands over the money, it is not on the grounds of his mistakenly trusting an untrustworthy repayment assurance. According to the defender, in both the mugging and false promising cases the description of the agent's end includes a description of how, through bringing about a particular mental state in the victim (fear or trust), he will realize a certain state of affairs (get money). The defender of the logical contradiction interpretation suggests that, in assessing whether an agent treats another merely as a means, his end must always be presented in this sort of way. But this effort to rescue the logical contradiction interpretation fails. An initial difficulty is that it is not logically impossible for end sharing to take place in either case as just described. For the sake of brevity, let us consider only the mugging case. In admittedly fanciful but logically possible circumstances, a passerby can intentionally contribute to realizing the end of a loiterer's obtaining $100 through getting her to fear that if she does not give it up, she will be harmed. She might, for example, hire a hypnotist to make her forget for a while that she is pursuing this end, to render her particularly anxious of being injured or killed by a gun, and to implant in her an urge to be in proximity to this particular loiterer. Assuming the hypnotist is successful—and it is surely logically possible that he is—when the passerby encounters the loiterer, he will obtain $100 from her through getting her to fear that if she does not give it up, she will be harmed. Moreover, that he obtains the money in this way will, in part, be her doing. Second, this effort to rescue the logical contradiction interpretation relies on a highly artificial characterization of agents' ends—for example, the characterization of the loiterer's end as that of "obtaining $100 through getting this passerby to fear that if she does not give it up, she will be harmed." It is far more natural to say that the loiterer has the end of obtaining $100 from this passerby and that the *means* to it that he plans to take is getting her to fear that if she does not give it up, she will be harmed. Imagine that before the loiterer threatens the passerby with his gun, she holds out $100 for him, which he takes from her hand. As the loiterer can surmise from her body-language and her dress, she is a very wealthy woman who is acting from charity, rather than from fear. If the complex description of the loiterer's end is on target, then we must say that he has not attained it. But it is plausible to imagine that the loiterer has attained his end, namely that of getting $100 from the passerby. The complex characterization of the false promisor's end is also artificial. We can easily imagine him thinking that he has attained his end, namely that of getting money from the lender, even if the latter's basis for giving it to him was not his mistakenly trusting an untrustworthy repayment assurance.

from *choosing* whether to contribute to the realization of that end or not."[20] The lender in our example cannot share the borrower's end of getting money without ever repaying it; for the borrower's false promise obscures his end and thus prevents the lender from choosing whether to contribute to it. Therefore, implies the interpretation, the borrower is in Kant's view treating the lender merely as a means.

This reading of possible end sharing has unacceptable normative implications. Consider a couple of young men hiking in the Rockies for the first time who find themselves on a mountain in late afternoon without provisions and unsure of the way down. To their relief, they spot another hiker, someone whom they saw park his car in the same area below: a burly man dressed in combat fatigues. They follow him, using his skill in choosing a path in order to get down the mountain safely. The young men realize that they could, but choose not to, tell the climber that they are following him. They stay far enough back so that they remain undetected; for they are embarrassed by their lack of self-reliance. The way they act prevents the man from choosing whether to contribute to the realization of their end. If the interpretation of Kant we are here considering is correct, then he is forced to embrace the following implausible view: Since the hiker cannot share the young men's end, they are treating him merely as a means and thereby acting wrongly.

One might respond that the young men's behavior does not really prevent the hiker from choosing whether to contribute to their end; it simply fails to facilitate his doing so. But if the men know, and it is reasonable to assume they would, that the hiker can find out what they are up to only if they tell him, then the way they act does for all practical purposes prevent him from choosing to contribute to their end. Moreover, imagine that one of the young men wants to tell the hiker that they are following him. But the other, who is more embarrassed by their ineptitude, prevents him from doing so. Of course, the other is, in effect, preventing the hiker from having a choice as to whether to contribute to the young men's end of getting down the mountain safely. However, it is implausible to claim that he is treating *the hiker* merely as a means and thereby acting wrongly, either in following the hiker or in preventing his friend from telling him of their use of him. Sometimes in pursuing an end an agent uses another without the other's being aware of his doing so. And, contrary to Kant's views (according to this interpretation), there need be nothing morally problematic about this lack of awareness, even when it is within the agent's power to eliminate it.

Perhaps we should take from the false promising passage a third interpretation of possible end sharing, *roughly* the view that the promisee cannot share the promisor's end in the sense that, in typical cases, it would be practically irrational for him to share this end. In typical cases, it would be irrational for the promisee to try to realize the end of making a loan that is never to be repaid. For this end's being brought about would

[20] Christine Korsgaard, *Creating the Kingdom of Ends* (Cambridge: Cambridge University Press, 1996), 139.

prevent him from attaining other ends he is pursuing, ends such as buying rose bushes, saving money for college, and, of course, just plain getting his money back.

The notion of irrationality at work here is familiar. In the *Groundwork*, Kant introduces a principle that Thomas Hill, Jr., calls "the hypothetical imperative": If you will an end, then you ought to will the means to it that are necessary and in your power, or give up willing the end.[21] To will an end is to set it for oneself and to try to realize it. Kant implies that the hypothetical imperative is a principle of reason: all of us are rationally compelled to abide by this principle. An agent would act contrary to the hypothetical imperative and thus irrationally by willing an end yet, at the same time, willing another end, the attainment of which would, he is aware, make it impossible for him to take the otherwise available and necessary means to his original end. If he willed this other end, he would, in effect, be failing to will the necessary means in his power to his original end. An agent would violate the hypothetical imperative, for example, by willing now to buy a car and yet, at the same time, willing to use the money he knows to be required for the down payment to make a gift to his nephew. If he willed to make the gift, he would, in effect, be failing to will the necessary means in his power to buy a car. The Kantian hypothetical imperative implies that it is irrational to will to be thwarted in attaining ends that one is pursuing. In typical cases, if a promisee willed the end of a false promisor, she would be doing just that.

It is important to recognize that there are two things that an agent who has willed something can do which would bring his action into compliance with the hypothetical imperative. He can either will the means that are necessary and in his power to the end (which, of course, would rule out his willing to be thwarted in attaining the end) or he can give up willing the end. For example, the hypothetical imperative would not imply that it was irrational for the person described above to cease willing now to buy a car and instead use the money that he knows to be required as a down payment on it to make a gift to his nephew.

We might say that a person cannot share an agent's end if and only if the person has an end such that his pursuing it at the same time that he was pursuing the agent's end would violate the hypothetical imperative, and he would be unwilling to give up his pursuit of this end, even if he was aware of the likely effects of the agent's pursuit of her end. Suppose that an agent plans to use a healthy person to obtain a heart and lungs for transplant, that is, to extract them from him in an operation that would kill him. We can imagine that the person has many ends, for example, that of attending his daughter's wedding. According to the hypothetical imperative, it would be irrational for him to pursue this end at the same time he was pursuing the end of the agent's getting from him a heart and lungs for transplant. The account implies that the person

---

[21] See Thomas Hill, Jr., *Dignity and Practical Reason in Kant's Moral Theory* (Ithaca: Cornell University Press, 1992), 17–37. Kant's main discussion of imperatives is in the *Groundwork*, especially 413–18. I give an overview of his views in Samuel Kerstein, "Imperatives: Categorical and Hypothetical," in the *International Encyclopedia of Ethics* (Oxford: Blackwell), 2012.

cannot share the agent's end just in case he would be unwilling to give up his end of attending his daughter's wedding against the background of an awareness of the likely effects of the agent's pursuit of his organs, including the loss of his own life, and, for example, the preservation of others' lives.

But this account suffers from a significant shortcoming. It would, let us assume, be practically irrational in the sense we have mentioned for the victim of a mugging to pursue the mugger's end of getting his money, for it would preclude him, say, from attaining his end of going to the game that evening. However, it is easy to imagine the victim being willing to cease pursuing his end of going to the game if he was aware of the likely effects of the agent's pursuit of her end, including that, in order to pursue his end, he would have to undergo the trauma of resisting a mugging. If the victim was willing to give up his end, then the account implies, implausibly, that he can share the mugger's end of getting his money.

Fortunately, it is not hard to refine the account so that it avoids this implication. A person cannot share an agent's end, let us say, if and only if:

The person has an end such that his pursuing it at the same time that he pursues the agent's end would violate the hypothetical imperative, and the person would be unwilling to give up pursuing this end, even if he was aware of the likely effects of the agent's pursuit of her end and aware that, based solely on his preference, the agent would give up her pursuit of her end.

Of course, the mugging victim would not give up his end of going to the game if he was aware that the mugger would act in accordance with his preference that she not mug him. So in this form, the account does not suffer from the shortcoming.

This notion of conditions under which a person cannot share an agent's end are included in the End Sharing Account:

An agent treats another merely as a means if it would be unreasonable for the agent to believe that the other can share the proximate end or ends the agent is pursuing in treating him as a means.

The notion of reasonableness at work here is non-moral. What it is reasonable for an agent to believe is what the evidence available to the agent favors, given the information he has, his education, his upbringing, and so forth. On this usage, it might be unreasonable for an agent to believe that another can (or cannot) share a particular end even if experts, say a group of leading psychologists, would believe the opposite. The account invokes an agent's proximate end or ends, namely what she aims to bring about directly from her use of the person. Her proximate end might also be her ultimate end, say, if she uses another for her own pleasure. But her proximate end might be far removed from her ultimate end. Someone might, for example, use another to develop her skill as a harpsichordist in order to earn a good living as a musician so she can put her little sisters through college, and so forth. The account invokes proximate ends because they are far more intimately connected to the use that brings them about than ultimate ends need be.

Unfortunately, the End Sharing Account has serious shortcomings. Suppose that Pete and Andre are competing in the men's singles final at a season-ending tennis tournament. Both players are going to retire after the tournament, and each player has as his goal to end his career by defeating the other. Each player's proximate end coincides with his ultimate end. Pete is treating Andre as a means; for he is intentionally doing something to Andre, that is, trying to beat him, in order to secure his goal, and he is relying on Andre's participation in order to secure it. But, according to the present account, Pete is also treating Andre *merely* as a means. It would not be reasonable for Pete to believe that Andre would be willing to give up his end if Andre was aware of the likely effects of the competition (e.g., a hard-fought match) or even if he was aware that Pete was willing to give up his pursuit of his end, if that is what Andre preferred. (We can easily imagine both competitors thinking: Why should I care that the other would give up his goal if I gave up mine. I don't want to give up mine!) Moreover, it is not reasonable for Pete to believe that Andre could pursue Pete's end of ending his career by defeating him (Andre) without practical irrationality. In pursuing this latter end, (e.g., by purposefully throwing the match) Andre would obviously be willing to be thwarted in attaining his own end. So on the present account, it turns out that Pete's treatment of Andre is (*pro tanto*) morally impermissible. In general terms, the End Sharing Account has the following implication. Suppose an agent is treating another as a means in a competition. It is, of course, typically going to be unreasonable for the agent to believe that in light of the likely effects of the agent's pursuit of his end, the competitor would give up his end, even if the competitor knew that the agent would cease trying to beat him if he (the competitor) wished it. Moreover, it is typically unreasonable for the agent to believe that his competitor can, rationally speaking, both pursue the agent's end and strive to secure his own. The End Sharing Account entails that competitive behavior of a common sort is *pro tanto* wrong.

One might reply that though this implication initially seems to discredit this account of treating others merely as means, reflection reveals otherwise. If Pete's end was not to have a lifetime winning record over Andre, but rather to develop his capacities as a tennis player, then he would not be treating Andre merely as a means. For it is reasonable for Pete to believe that this is an end that Andre can share. (Even if Andre himself was playing to get the overall edge over Pete, he could, without practical irrationality, pursue at the same time the end of developing Pete's skills. After all, in the course of losing Pete might grow as a tennis player.) Competitors often aim ultimately at developing their own capacities, rather than at winning. When they do strive to be the sole victors in a competition, it is not implausible to hold that they are treating their opponents merely as means and are thus acting *pro tanto* wrongly, concludes the reply.

In my view, Pete and Andre would in some sense be more virtuous if each could share the other's end. Competitors who each have as an end to develop their own capacities seem morally more attractive than ones who each have as an end to best their rivals. There is something admirable in holding that, ultimately, one is "competing against" oneself. However, I think that most of us would find unacceptably strong the judgment that it is (*pro tanto*) *morally wrong* to act as Andre and Pete do in the original example. Suppose, as most of us believe, that their actions are not wrong, all things considered. If their actions are nevertheless *pro tanto* wrong, then Pete and Andre must each have some reason to try to defeat the other, which reason outweighs his strong reason to refrain from treating the other merely as a means. But what would that reason be? In sum, some competitors do aim ultimately at improving their own skills, but others (and, it seems to me, a lot of others) aim ultimately to be victorious. To conclude that these others are thereby treating their opponents in a (*pro tanto*) morally impermissible way seems far-fetched.

Competition cases are not the only ones regarding which, according to many of us, the End Sharing Account yields counterintuitive results. Suppose that a movie director aims to have Ronaldo play a supporting role in an upcoming production. She wants to use him in that part. As Ronaldo makes clear to her, he really wants the lead role. The director tells him firmly and repeatedly that she has already found someone for that part and that she will hire him for the supporting role or not at all. Ronaldo insists that the director will change her mind when she sees what he can do at rehearsal. But, in full awareness of the director's terms, he signs on to play the supporting role. It would presumably be unreasonable for the director to believe that Ronaldo can share her aim of having him in the supporting role, for it might be reasonable for her to believe the following: Ronaldo's pursuing this aim would amount to his trying to be thwarted in realizing his own end of playing the lead. After all, an actor cannot play these two roles at once. Moreover, Ronaldo would not give up his end of playing the lead in light of an awareness of the likely effects of his undertaking the supporting role or in light of willingness on the director's part to give up on using him in the supporting role. So, according to the End Sharing Account, the director would be treating Ronaldo merely as a means and thus acting *pro tanto* wrongly. But that conclusion seems implausible. The director is not doing what Ronaldo wants, but she is not "just using" him either. A person might have the power to determine whether an agent uses him and agree voluntarily to this use, even though he cannot share the agent's (proximate) aim. In some of these cases, it is far-fetched to claim that the agent is treating the person merely as a means.

In the false promising passage from the *Groundwork*, Kant suggests the idea that an agent uses another merely as a means if in his use of the other he is pursuing an end that the other cannot "contain in himself," that is, share. We have explored three interpretations of what the other's being unable to share an agent's end amounts to: it is logically impossible for the other to share his end; the agent has rendered him unable to choose whether to contribute to it; and the other's willing the agent's end would violate the hypothetical imperative. (This last interpretation is embedded in the End Sharing Account.) On each of these interpretations, Kant's idea that an agent uses another merely as a means if in his use of the other he is pursuing an end that the other cannot share turns out to be problematic. So it makes sense to look elsewhere for an interpretation (or reconstruction) of Kant's false promising passage.

3.3 Rational Consent

According to Kant, the person whom the agent is using by making a false promise "cannot possibly agree" to the agent's use of him. One might take this to mean that the person cannot *rationally* consent to this use. According to the Rational Consent Account:

> Suppose an agent uses another. She uses him merely as a means if the other cannot rationally consent to her use of him.

But what does it mean for someone to be able, rationally speaking, to consent to someone's using him? In the absence of an answer to this question, the account is too vague to be useful.

Fortunately, Derek Parfit has recently offered an answer. Largely on the basis of reflection on Kant's false promising passage, Parfit contends that Kant embraces the "*Consent Principle*," which says that "it is wrong to treat anyone in any way to which this person could not rationally consent."[22] We will flesh out the Rational Consent Account with the help of Parfit's notion of when someone can, rationally speaking, consent to being used.[23] However, Parfit does not himself attribute to Kant or himself defend the Rational Consent Account of treating others merely as means.[24] Parfit develops a *different* view of treating others merely as means, a view that we discussed briefly at the beginning of the chapter.

In order to facilitate our exploration of when, according to Parfit, someone can rationally consent to being treated as a means, let us make some simplifying assumptions. Let us suppose that the people who might be used are informed. They understand how they might be used, that is, what will be done to them and to what purpose, as well as the effects the use will have. Let us also assume that those who might be used have the power to give (or to withhold) consent in what Parfit calls the "act-affecting sense."[25] When we ask whether they can rationally consent to being used, we are asking whether it would be rational for them to consent (or dissent) supposing that their choice would determine whether or not they were used.

Against the background of these assumptions, we can say that, according to Parfit, a person can rationally consent to being treated as a means just in case he has *sufficient*

---

[22]   Parfit, *On What Matters*, 181.

[23]   According to Parfit, an agent's treating someone in some way is a broader notion that her using him: "We should be counted as *treating* people in some way when we know that our act, or one of its possible alternatives, would or might affect these people in some way, or be an act with which they would have some personal reason to be concerned" (Parfit, *On What Matters,* 184). All cases of an agent's being able to rationally consent to being used in some particular way would, of course, count as cases of his being able to consent to being treated in that way.

[24]   That Parfit attributes the Consent Principle to Kant, but not the Rational Consent Account to him is odd. For it is in describing what treating another merely as a means amounts to that Kant writes in terms that inspire Parfit to attribute the Consent Principle to him.

[25]   Parfit, *On What Matters*, 183–4.

reasons to consent to it. Parfit embraces an "objective" view of reasons. He holds that "there are certain facts that give us reasons both to have certain desires and aims, and to do whatever might achieve these aims. These reasons are given by facts about the *objects* of these desires or aims, or what we might want or try to achieve."[26] So, for example, the fact that a child is being hurt by the splinter stuck in her hand gives me reason to want to remove it and to try to do so.[27] According to Parfit, we have impartial as well as partial reasons for consenting to be treated in various ways. Our impartial reasons are "person-neutral."[28] We need not refer to ourselves when we describe the facts that give rise to these reasons. For example, the fact that some event would cause great suffering to a particular person gives us reason (albeit perhaps not sufficient reason) to prevent the event or relieve the suffering "whoever this person may be, and whatever this person's relation to us," says Parfit.[29] Our partial reasons are "person-relative": they "are provided by facts whose description must refer to us."[30] The fact that the little girl being hurt by the splinter in her hand is *my* daughter gives me a partial reason to pull it out. We each have partial reasons to be particularly attentive to our own well-being and to the well-being of those in our circle, for example, our family and friends, according to Parfit. He holds the following "wide value-based objective" view of sufficient reasons for acting: "When one of our two possible acts would make things go in some way that would be impartially better, but the other act would make things go better either for ourselves or for those to whom we have close ties, we often have sufficient reasons to act in either of these ways."[31] For example, regarding a case in which a person could either save himself from some injury or do something that would save some stranger's life in a distant land, Parfit says that the person has sufficient reasons to do either one. In a similar vein, a person can have sufficient reason to consent to being treated as a means by virtue of some impartial reason, such as the fact that his being so treated will save many lives, even if he also has sufficient reason to dissent from being treated as a means by virtue of some partial reason, such as the fact that his being so treated will result in suffering for him.

The Rational Consent Account holds that an agent uses another merely as a means if the other cannot rationally consent to her use of him. Parfit would presumably reject the Rational Consent Account. As we have noted, he champions a different understanding of treating others merely as means. Nevertheless, we can employ the notion of rational consent that he develops in an effort to make the Rational Consent Account determinate. When specified with this notion, the account holds that an agent

[26] ibid. 45.
[27] Parfit contrasts an objective view of reasons with a subjective view, according to which our reasons for acting are "all provided by, or depend upon, certain facts about what would fulfill or achieve our . . . desires or aims" (*On What Matters,* 45). On this view, I would presumably have a reason to remove the splinter from the child's hand because I want (or would want, if I considered all of the facts) her not to be in pain.
[28] Parfit, *On What Matters,* 138.
[29] ibid.
[30] ibid.
[31] ibid. 137.

uses another merely as a means if the other does not have sufficient reasons (whether they be partial or impartial) to consent to her use of him. (Again, we are assuming that the other is well-informed about the nature and effects of the agent's using him, and that his withholding his consent from her doing so would lead her to refrain from using him.)

There are several difficulties with the Rational Consent Account. One problem is that it fails to designate as such what many of us take to be particularly clear instances of an agent's treating another merely as a means. Parfit discusses a familiar case in which a train is headed for five innocent people.[32] One person is on a bridge above the track. The only way to save the five would be to open, by remote control, the trap-door on which the person is standing, so that he would fall in front of the train. He would be killed, but he would trigger the train's automatic brake. If a bystander opens the trapdoor, then she uses the person on the bridge as a means to save the five. Intuitively speaking, many, indeed, I believe, most of us, would say that she uses him *merely* as a means. But the Rational Consent Account, at least interpreted in accordance with Parfit's own notion of when someone has sufficient reason to consent to being treated in some way, fails to generate this verdict. According to Parfit, the person on the bridge has sufficient reason to refrain from consenting to being used to stop the train. After all, it will result in something bad for him, namely his premature death. But he also has sufficient reason to consent to being used to stop the train, for his being used in this way would save the lives of five innocent people, contributing to an outcome which Parfit presumably takes to be impartially best.[33] So the Rational Consent Account does not help us to reach the conclusion that the bystander treats the person on the bridge merely as a means. Of course, the account does not block this conclusion either. For all that the account implies, the bystander may well be treating the person on the bridge merely as a means. The account purports to offer a sufficient, not a necessary, condition for an agent's treating another in this way.

A second difficulty with the account emerges when we consider how we reached the conclusion that it does not imply that the bystander was treating anyone merely as a means. We had to appeal to what the person on the bridge had sufficient reason to consent to. But the notion of what someone has sufficient reason to consent to is hard to specify. Indeed, it seems every bit as difficult to elaborate as the notion of treating others merely as means. So it is questionable whether an account that appeals to what someone has sufficient reason to consent to can do much to clarify the Mere Means Principle.

Returning to the case at hand, Parfit claims that the person on the bridge did indeed have sufficient reason to consent to being used as a train-decelerator. But many of us doubt that. Adding some detail to the case, let us imagine that the person on the bridge is aware of what is going on and is asking the bystander not to trigger the trapdoor.

---

[32] ibid. 218.   [33] ibid. 220.

According to Parfit, this refusal of the person to consent to being used does not change the fact that he has sufficient reason to consent.[34] However, many of us believe that although he has some impartial reason to consent to being used (e.g., since overall well-being would thereby best be promoted), he does not have sufficient reason to consent. Remaining within Parfit's general way of thinking about reasons, we might say that in this case the person's partial (i.e., personal) reasons not to consent to being used as a train-decelerator are so much weightier than his impartial reasons for consenting that the former render the latter insufficient.[35] Among the agent's partial reasons not to consent might not be just the fact that if he gets used in this way he loses out on the rest of his life, but also the fact that he has made an informed choice not to be used in this way, even though it is necessary in order for five others to survive.

In general, one's having made an informed choice not to do something can itself contribute to one's personal reasons not to do it, many of us hold. Consider two scenarios. In the first, I have a particular personal reason not to undertake a rescue: doing so will result in my losing a leg. But I have made an informed decision to undertake the rescue anyway. In the second scenario, I have this same particular personal reason not to undertake a rescue. And I have made an informed decision *not* to undertake one. My having made this decision adds to the overall partial reason I possess not to go ahead with the rescue, many of us believe: I have *more* partial reason not to go ahead with the rescue in the second scenario than I do in the first.

My aim here is not to settle thorny questions concerning the (in)sufficiency of the person's partial or impartial reasons to consent to being used in the bridge case. My aim is rather to underscore that there are such questions. At the beginning of the chapter I suggested as a methodological guideline that we try to avoid accounts of treating others merely as means that rely on notions that are just as difficult to pin down as the one(s) we are trying to clarify. The Rational Consent Account does not do a good job of respecting that guideline, for, as we have seen, in order to apply it we must appeal to the elusive notion of what people have sufficient reason to do.

A third problem with the account is the most serious. It implies that an agent is treating another merely as a means in cases in which, intuitively speaking, she is doing no such thing. Suppose, for example, that an executive hires a personal assistant, using him to increase her productivity. She has in no way tried to deceive him regarding any aspect of the job or to force him to work for her. He voluntarily accepts the position. He has some reason to take the job; it will make him happy in the short run. But, as he is aware, he has far more reason to take another position, namely one as a teacher in a high school. He has strong partial and impartial reasons for becoming a teacher.

---

[34] ibid. 220–1.

[35] Parfit himself embraces the possibility that one sort of reason for acting (or, presumably, consenting to be treated in some way), for example, a partial reason, can override another reason, for example, an impartial reason, so that although one does have a reason of the latter sort for acting, he does not have sufficient reason for doing so. See *On What Matters*, 140.

Although it will be difficult for him at first, he will thrive in the long run, and he will help many students lead better lives than they otherwise would. He has *so much more* reason to take the job as a teacher that he does not have sufficient reason to become a personal assistant. According to the Rational Consent Account, in employing the job seeker, the executive is treating him merely as a means. But this result is implausible, to say the least. Not all cases in which another fails to have sufficient reasons to consent to an agent's using her are cases of the agent's using her merely as a means. There are cases in which a person gives his informed consent to be used by an agent and in which that consent does not at all result from the agent's having coerced or deceived him. In some of those cases, like the one just described, the agent is not using the person merely as a means.

3.4 Possible Consent

We are still searching for a plausible sufficient condition for an agent's treating another merely as a means. Accounts featuring notions of end-sharing and rational consent have proved inadequate. But Onora O'Neill suggests that Kant might endorse a different account: An agent uses another merely as a means if the other cannot consent to her use of him.[36,37] If it is not possible for an agent to dissent from being used, implies O'Neill, then he cannot consent to it. It is possible in the relevant sense for someone to dissent from being used, she suggests, only if he can avert this use by withholding his consent to it.[38] So an agent cannot consent to being treated as a means if he does not have the ability to avert his being treated as such by withholding his agreement to it. According to O'Neill, if an agent deceives or coerces another, then the other's dissent is "in principle ruled out," and thus so is his consent.[39] Suppose, for example, that an auto mechanic tricks a customer into authorizing an expensive repair. The customer does not really have the opportunity to dissent to the mechanic's action by refusing to give his consent to it. For he does not know what her action is, namely one of lying to him about what is really wrong with his car. (If he did know what her action was, then he wouldn't be deceived.) Or, moving to another example, suppose that a mugger approaches you on a dark street, points a gun at your torso, and tells you with chilling candor that unless you give him all of your money, he will kill you. He leaves you no

---

[36]   O'Neill suggests that in Kant's view an agent's using another merely as a means amounts to his acting *on a maxim* to which the other cannot consent (Onora O'Neill, *Constructions of Reason* (Cambridge: Cambridge University Press, 1989), 113). In a departure from O'Neill's presentation, the Possible Consent Account does not invoke Kant's notion of a maxim. It is notoriously difficult to specify what Kant means by a maxim, and for the sake of simplicity I do not wish to invoke maxims here. So far as I can tell, this departure from strict Kantianism does not affect the substance of what follows.

[37]   Korsgaard seems to agree with O'Neill on this point. "The question whether another can assent to your way of acting," she writes, "can serve as a criterion for judging whether you are treating her as a mere means" (Korsgaard, *Creating the Kingdom of Ends*, 139).

[38]   O'Neill, *Constructions of Reason*, 110.

[39]   ibid. 111.

opportunity to avert his use of you by withholding your consent. So in O'Neill's sense, you cannot consent to his action. The mugger is treating you merely as a means.[40] In order for a person to be able to avert an agent's use of him, the person must have a basic understanding of what that use is. The person's having such an understanding presumably involves his being aware of how the agent wishes to use him (e.g., by taking a sample of his blood), what the likely effects on him will be (e.g., it might hurt), and what the agent aims to use him for (e.g., to gain information on the effectiveness of a drug).

Is it a strength of this account of treating another merely as a means that, contrary to the End Sharing Account, it does not imply that one need be doing so in striving to be the sole victor in a competition? Again, suppose that Pete is trying to defeat Andre in an effort to end his career with a victory over him. At the time Pete's action takes place, Andre probably cannot avert it by withholding his consent. (A proclamation by Andre such as "I refuse to allow you to try to beat me" would not only be odd, but would also likely be ignored by his opponent.) So it appears that the account also implies that Pete is doing something wrong.

But it is, I believe, in the account's spirit to notice here that Andre has been able to consent to the competitive order inherent in the world of professional tennis. It has been open to him to remove himself from this order by withholding his agreement to participate in it. He has had the ability to avert his being subject to rules according to which it is obviously legitimate and anticipated that his opponents on the court will treat him as a means to attaining their own competitive ends. I believe that on O'Neill's considered view, Andre's possession of this ability entails that he can consent to Pete's use of him. If an agent can consent to a set of rules according to which a particular kind of use of him is obviously both legitimate and anticipated, then, for purposes of the account, he counts as being able to consent to the use itself.

Although the account allows us to avoid the implausible conclusion that it is morally impermissible to pursue in a competition the goal of defeating another who is also striving for victory, it suffers from two difficulties. An initial problem is easy to illustrate with an example. Suppose that I hail a cab and ask the driver to take me to the airport. But, unknown to me, the driver cannot refuse my request; he cannot avert my use of him. For he has been hypnotized into being unshakably convinced that he cannot turn

---

[40] O'Neill actually says that it is possible for someone to dissent from a course of action only if he "can avert *or modify* the action by withholding consent and collaboration" (*Constructions of Reason*, 110, italics mine). On one natural interpretation, this implies that if a person can modify an agent's use of him, then he can dissent to it. But this implication seems problematic, as a slightly altered version of the mugging case will show. Suppose that the mugger tells the victim that unless he gives him all of his money, he will kill him. But the victim comes back and says that he will give him half of his money without a struggle and the mugger accepts the arrangement. The victim was able to modify the mugger's use of him by withholding his agreement to it. So, it seems, O'Neill is implying that the victim can dissent from the mugger's use of him. But that seems counterintuitive. In any case, it does not seem to be O'Neill's considered view. For, as we have seen, she also says that in cases of coercion, which the one just described surely is, dissent is in principle ruled out.

down any request I make. In such a scenario, although the driver cannot avert my use of him by withholding consent, it would be implausible to think that I have acted wrongly in my use of him. It is not I, but rather the hypnotist who has (presumably) acted wrongly in his use of the driver. It would be easy to multiply cases such as this in which an agent is unaware that another cannot avert the agent's treatment of him by withholding his agreement to it but in which it seems implausible to conclude that the agent is treating the other (*pro tanto*) wrongly. In such cases, the agent reasonably believes that the other can dissent from being used. Here again reasonable belief is an epistemic rather than a moral notion.

There is another serious difficulty with the account on the table. It does not suffice for an agent to treat another merely as a means that the other simply *be* unable to consent to the way he is being used. If it did suffice, then a passerby giving cardiopulmonary resuscitation (CPR) to a collapsed jogger in order to impress his girlfriend would be treating the jogger merely as a means and thus acting (*pro tanto*) wrongly. But the passerby does not seem to be doing anything that is (*pro tanto*) morally impermissible.

In light of these difficulties, let us consider the following, more sophisticated account, which I call the Possible Consent Account:

Suppose an agent uses another. She uses him merely as a means if it is reasonable for her to believe that something she has done or is doing to the other renders him unable to consent to her using him.

Of course, although the collapsed jogger has no opportunity to consent to the passerby's giving him CPR, the passerby has not put him in that position. So this account avoids the unwelcome implication that the passerby treats the collapsed jogger merely as a means.

However, the account does have serious shortcomings. First, the Possible Consent Account fails to designate as such some cases that we, intuitively speaking, would surely classify as treating others merely as means. Suppose, for example, that a geneticist has travelled to a remote village to gather blood samples for her research. But she has found no one who wishes to give one. For the villagers believe themselves to have been exploited by researchers in the past and wish to give them no further help. Before departing on her long trip home, the geneticist visits a clinic where she notices a patient who has just emerged from an operation and is unconscious. She sneaks a blood sample from him and leaves. According to reflective common sense, the geneticist treats the patient merely as a means. But the Possible Consent Account does not itself enable us to draw that conclusion. The patient is unable to consent to her use of him, that is, he has no opportunity to avert her action by withholding his agreement to it. But the geneticist has not rendered him unable to consent; it was not she who put him under. It would be easy to multiply such cases. Think for example of a case where one person knocks someone out with a "date rape" drug. Another person, who had no knowledge of or involvement in drugging the victim, sexually assaults her. These cases are a subset of ones in which someone cannot consent to an agent's using him in some way, but in which his being unable to consent is not a result of anything that the agent himself has done.

As it stands, the Possible Consent Account not only fails to capture some cases of an agent's treating another merely as a means, but, more disturbingly, it also designates as such some cases that, intuitively speaking, are not. For example, in order to make your spouse's birthday party a surprise for her, you need to lie to your sister-in-law about your whereabouts during a certain afternoon. You use her to quell your spouse's suspicions regarding your plans. As you realize, if you told your sister-in-law about the party, she would be unable to keep the secret from your spouse. According to the Possible Consent Account, you treat your sister-in-law merely as a means. For your deception leaves her with no opportunity to avert your use of her. But this conclusion seems counterintuitive. Granted, the conclusion implies that your action is *pro tanto* wrong, but not necessarily wrong, all things considered. Yet suppose we embrace the plausible idea that your action is not wrong, all things considered. If we cling to the Possible Consent Account, we must then hold that you have some reason to treat your sister-in-law merely as a means that outweighs your strong reason not to treat her in this way. But you do not seem to have such a reason: the pleasure your spouse will experience at a surprise party (as opposed to a party she foresaw) does not seem to provide such a weighty reason. The Possible Consent Account seems not to permit us to hold that your use of your sister-in-law was morally permissible.

Here is another case of what we intuitively think of as morally permissible deception. Suppose that, in order to save the life of an innocent witness to a crime, you use her to pass on a lie you have told her to the perpetrator, Brown. If Brown didn't believe the lie, he would kill the witness.[41] You realize that if you let the witness in on what was necessary to save her life and told her to lie to Brown herself, she wouldn't be able to do so effectively. Your treatment of the person renders impossible her consent to your use of her. But it is implausible to conclude that you are treating her merely as a means.

In these two cases, it is reasonable to think that the person you are using can share your ends: your sister-in-law can share the end of your spouse not getting suspicious regarding a surprise party, and, of course, the witness can share the end of Brown's coming to believe some lie. And that, I venture, is why the person's inability to consent to your use of her falls short of plausibly implying that you are using her merely as a means.

3.5 The Hybrid Account

Thus far we have found inadequate several accounts of treating others merely as means, including the End Sharing, Rational Consent, and Possible Consent accounts. Since we considered a single specification of the Rational Consent Account, we have

---

[41] Parfit, *On What Matters*, 178.

obviously not ruled out the possibility that someone might develop a plausible version of it. I suspect, however, that any version of the account that both identified central cases of a person's treating another merely as a means (e.g., someone's forcing one person into a fatal fall off of a bridge as a means of saving five) and avoided implausible implications (e.g., that the executive just uses her new personal assistant) would need to appeal to normative concepts every bit as obscure and contested as that of the notion of treating others merely as means itself.

The Possible Consent and End Sharing Accounts suffer from serious shortcomings, but they might be combined into an account that is both plausible and clarificatory. Let us explore a sufficient condition for an agent's treating another merely as a means that incorporates both the notion, taken from the Possible Consent Account, that the other cannot consent to the agent's use of him, and the notion, taken from the End Sharing Account, that he cannot share the end she is pursuing in using him. According to the Hybrid Account:

> Suppose an agent uses another. She uses him merely as a means if it is reasonable for her to believe both that he is unable to consent to her using him and that he cannot share the proximate end(s) she is pursuing in using him.

The Hybrid Account incorporates the non-moral notion of reasonable belief we have highlighted in the two earlier accounts. Does the Hybrid Account remain true to our methodological imperative to minimize appeals to controversial moral concepts, yet avoid the problems that plague the accounts from which it stems?

Consider first the Possible Consent Account. It did not enable us to reach the conclusion that the researcher who, without permission, took a blood sample from an unconscious patient was treating him merely as a means. But the Hybrid Account generates this conclusion. For it is reasonable for the researcher to believe both that the patient is unable to consent to her use of him and that he cannot share the end she is pursuing in using him, namely that of getting blood for her research. Moreover, unlike the Possible Consent Account, the Hybrid Account is free from implausible implications in cases of what we take to be permissible deception. The account does not imply, for example, that in lying to your sister-in-law in order to make your spouse's birthday a surprise, you are treating her merely as a means. It is presumably reasonable for you to believe that she can share your end of making your spouse's party a surprise. If it is *not* reasonable for you to believe this, say because, as you know, your sister-in-law thinks that your spouse would hate a surprise party, then it seems plausible to conclude that you would be treating your sister in law merely as a means in using her to deceive your spouse.

The Hybrid Account also steers clear of the pitfalls to which the End Sharing Account succumbs. It does not entail that competition of a common sort, such as that between Pete and Andre, involves opponents treating one another merely as means. Each competitor can consent to the other's use of him, for each has consented to a set of rules according to which the other's use of him is obviously both legitimate and anticipated. In addition, the Hybrid Account is consistent with the idea that if it is reasonable for an agent to believe that another can consent to her use of him, she might not be treating him merely as a means, even if it is also reasonable for her to believe that the other cannot share her (proximate) aim in using him. The Hybrid Account does not generate the dubious result that the film director is treating Ronaldo merely as a means.

But the Hybrid Account faces challenges. In a case that we will call "Arrest," white supremacists are planning an attack on a church in order to promote their racist ideology. A police officer enters a supremacist's house, gun drawn, and arrests him for questioning, treating him as a means of getting information about the planned attack. It is reasonable for the officer to believe that the supremacist cannot consent to his arrest, that is, that he cannot alter her action of taking him into custody by withholding his consent to it. Moreover, it is reasonable for the officer to believe that the supremacist cannot share the end she is pursuing in arresting him. For in willing the officer's proximate end of her getting information about the attack, the supremacist would, in effect, be willing to be thwarted in his end of carrying it out. And, it is reasonable for the officer to believe, that end is not one the supremacist would abandon if he was aware of the likely effects of the officer's pursuit of her end or even that, based solely on his preference that she do so, the officer would stop pursuing information about the attack. As the officer realizes, the supremacist is unyielding in his desire that the attack take place. So the Hybrid Account implies that the officer is treating the white supremacist merely as a means. This result is implausible, one might object.

One might be tempted to reply that, appearances notwithstanding, the officer is not treating the supremacist merely as a means. For in one respect, this case is like that of Pete and Andre described above. In much the same way as Andre has been able to consent to the competitive order of professional tennis, according to which it is obviously legitimate for opponents such as Pete to use him as a means towards their competitive ends, so the supremacist has been able to consent to our legal order, according to which it is obviously legitimate for police officers to use those plotting acts such as this in efforts to thwart them. Andre was able to avert his being subject to the rules of professional tennis by choosing not to play professionally. The supremacist was able to avert his being subject to our legal order by choosing to live somewhere else, ends the reply.

But the reply does not strike me as promising. It is one thing for Andre to have been able to opt out of becoming a professional tennis player. It is quite another for the supremacist to have been able to emigrate. What if he had no opportunity to do so? What if, for example, no other country was willing to accept him as a result of a combination of his being infected with the HIV virus at birth and being financially unable to support himself? We would in that case have to conclude that the officer was treating the supremacist merely as a means.

A more promising response is to take note of the following. An orthodox Kantian who embraced the Mere Means Principle (and the Hybrid Account as a specification of it) might hold that treating another merely as a means is always wrong, all things considered, and thereby be forced into the conclusion that the officer's action was morally impermissible. That conclusion would indeed be implausible, many of us believe. But we have a different take on the Mere Means Principle: while treating others merely as means is always wrong *pro tanto*, it can be morally permissible, all things considered. I offer an explanation below of why the officer's treating the supremacist merely as a means in this case (as well as in similar ones of what many of us take to be justified coercion) does not generate the conclusion that he is acting wrongly, all things considered.

Although in defending the Hybrid Account we need not conclude that the officer's action is wrong, all things considered, we do need to maintain that he is treating the supremacist merely as a means and thereby acting *pro tanto* wrongly. But that conclusion seems reasonable.[42] Several actions that police officers take, such as striking, detaining, or imprisoning people, are *pro tanto* wrong, it is safe to say. We are especially concerned that police officers, as opposed, for example, to haberdashers, have good character and good training, partly because police officers but not haberdashers have moral and legal license to do things that are normally wrong. So it does not seem incongruous to say that the officer's action in Arrest is *pro tanto* wrong. Nor, upon reflection, does it seem odd to conclude that the officer is treating the supremacist merely as a means. After all, as she realizes, she is using him to thwart one of his projects, without giving him the least opportunity to prevent her use.

Another case that, one might claim, casts a shadow of implausibility on the Hybrid Account is that of Attack. Suppose that two of the white-supremacists elude arrest, and one of them starts shooting at people in the church. A congregant pushes the other attacker into the shooter, in an effort to gain time for other congregants to escape. Let us assume that it is reasonable for the congregant to believe neither that the attacker whom he pushes can consent to his use of him, nor that he can share the end he is pursuing in using him. The Hybrid Account then entails that the congregant is treating one of the supremacists merely as a means and thus acting (*pro tanto*) wrongly. This conclusion is dubious, one might claim.

But is it? The Hybrid Account does not entail that the congregant is acting wrongly, all things considered. Indeed, I offer an account below of why the congregant's treating the attacker whom he pushes merely as a means does not render his action morally impermissible in this case or similar ones. The conclusion might seem implausible to some because they are reluctant to accept the idea that the congregant is treating the attacker merely as a means. But it makes sense to think of him as doing just that. For, as the congregant understands, he is giving the attacker no option to avert his use of him, and he is using the attacker for a purpose he cannot share. Upon reflection, our attitude towards treating others merely as means is akin to our attitude towards deceiving

---

[42] That was not my view when I initially considered the issue. See Samuel Kerstein, "Treating Others Merely as Means," 175–6.

THE MERE MEANS PRINCIPLE 79

others, I believe. We consider both to be wrong *pro tanto*, but not necessarily wrong all things considered.

Attack and Arrest are, respectively, cases of what we intuitively think of as morally permissible coercion. I have asserted that in both cases an agent treats another merely as a means, but that her doing so does not render her action morally impermissible. Why not? To put the point roughly, it seems to be because the persons whom the officer and the congregant are treating merely as means are themselves treating others merely as means. But let us put the point more precisely. A Wrong-Making Insufficiency Principle states:

Suppose that an agent, A, treats another, B, merely as a means. A's doing this does not typically suffice to render her action morally impermissible, if the following obtains: It is reasonable for A to believe that B is, rationally speaking, prevented from sharing the (proximate) end she is pursuing in using him as a result of his using someone else, C, and C can neither consent to this use, nor share the (proximate) end B is pursuing in using him.[43]

The notion of treating others merely as a means employed here is limited to that specified in the Hybrid Account. In a now familiar usage, "rationally speaking" is meant to invoke a requirement of prudential rationality discussed earlier, namely that expressed by the hypothetical imperative. This Wrong-Making Insufficiency Principle purports to specify one set of conditions under which one's *pro tanto* reason not to treat another merely as a means does not typically render it morally impermissible to do so.

When stated precisely, the principle seems unwieldy. But it is not difficult to apply. In Arrest, let us recall, a police officer takes a white-supremacist gang member into custody, thereby using him to get information about an upcoming attack on innocent people. The Hybrid Account entails that the officer is treating the supremacist merely as a means. But it is reasonable for the officer to believe that, in terms of prudential rationality, the white-supremacist is precluded from being able to share her end of getting information about the planned attack as a result of his commitment to the end of getting "revenge" on a minority group. It would violate the hypothetical imperative for him both to pursue this end and to will the officer's end. It is obviously also reasonable for the officer to believe that the white-supremacist knows full well that he is pursuing an end that members of the minority group cannot share and in a way to which they cannot consent. The conditions specified in our principle are realized. Perhaps that explains at least in part why we do not hold the officer's treating the supremacist merely as a means to be wrong.

In Attack, a congregant uses one person attacking his church to buy time for fellow congregants to escape. The congregant treats the attacker merely as a means. But it is

---

[43] For the sake of ease of expression, I have simplified the condition specified in this principle. Strictly speaking, it should read: It is reasonable for A to believe that B is, rationally speaking, prevented from sharing the (proximate) end she is pursuing in using him as a result of his using (*or having used or being about to use*) someone else, C, and C can (*or could or will be able to*) neither consent to this use, nor share the (proximate) end B is (*was or will be*) pursuing in using him.

reasonable for the congregant to believe the following: first, it would be irrational for the attacker to share his end, since he (the attacker) is himself using the congregants in pursuing "revenge," and, second, the other congregants can neither consent to the attacker's use of them, nor share the end he is pursuing in using them. Again, the conditions specified by our Wrong-Making Insufficiency Principle are fulfilled. And that might account to a large extent for our unwillingness to conclude that the congregant's use of the attacker is morally impermissible.

Two points are important to keep in mind. Our Wrong-Making Insufficiency Principle does not purport to be absolute. It is compatible with the observation that in some, presumably rare, cases, the principle's conditions are fulfilled, and yet it is nevertheless plausible to conclude that an agent's treating another merely as a means suffices to make his action wrong.[44] Moreover, this principle does not purport to capture all circumstances in which treating another merely as a means is morally permissible, all things considered. In Chapter 5 (5.6), we consider further circumstances in which it is.

3.6 Parfitian Objections to the Hybrid Account

Let us return to the bridge example. A passerby sees that a runaway train is about to kill five people on the tracks. As a means of saving them, she triggers a trapdoor, through which a pedestrian on the bridge above the tracks falls, slowing the train down enough so that its emergency brake kicks in. As the passerby foresees, the impact with the train kills the person. We can easily imagine a scenario in which the passerby would be treating the person on the bridge merely as a means, according to the Hybrid Account. The passerby is obviously using the person. She is intentionally doing something to him, namely making him drop through the trapdoor, in order to attain some end, namely that he stop the train. Of course, the passerby also intends the person to contribute to the realization of this end. Moreover, it is reasonable for her to believe both that the person cannot consent to this use and that he cannot share her end in using him.

Parfit would object to the Hybrid Account's implication that the passerby is treating the person on the bridge merely as a means. He embraces the view that, although in

---

[44] The following might be such a case: A man and a woman suspected of planning a crime are together with a detective in an interview room. The male suspect seizes the detective's weapon and forces him to reveal how much the police know about how the crime is supposed to unfold. As the suspect realizes, the detective cannot, rationally speaking, share his end of gaining this information. For if he gained it, the detective would be unable to deceive the female suspect, who knows a lot more about the planned crime, into revealing information that would prevent it. In this case, the conditions contained in our Wrong-Making Insufficiency Principle are fulfilled. The male suspect is treating the detective merely as a means. Moreover, it is reasonable for him to believe that the detective is, rationally speaking, prevented from sharing the end he is pursuing in using him because he, the detective, is using someone else, namely the female suspect, in pursuit of an end that she cannot share and in a way to which she cannot consent. Yet it seems that the suspect's treating the detective merely as a means itself suffices to make his action wrong.

triggering the trapdoor she might be acting wrongly, if she is, it would not be even in part because she is treating him merely as a means. Parfit suggests two grounds for his conclusion, neither of which justifies it, in my view.

First, he suggests that the passerby is not treating the person on the bridge merely as a means if her action is "governed by" the Consent Principle. Her action is governed by this principle, it seems, just in case she *tries successfully* to make it such that it does not involve treating anyone in any way to which this person fails to have sufficient reason to consent. Parfit believes that the passerby's action of triggering the trapdoor is governed by this principle. But that is doubtful. Although she might believe that the pedestrian has sufficient reason to consent to her use of him and, in general, *try to* limit her use of others to actions to which they have sufficient reason to consent, it seems to many of us that in this case she fails. For, as we noted above, it seems to many of us that the pedestrian does not in fact have sufficient reason to consent to being used to stop the train.[45]

Moreover, even if the pedestrian does have sufficient reason, in Parfit's sense, to consent, and the passerby tries to treat people only in ways to which they have such reason to consent, it does not follow that, intuitively speaking, the passerby avoids treating him merely as a means. Compare the well-known example in which there are five patients at a hospital in immediate need of different organs. One patient needs a kidney, another needs a heart, a third needs a lobe of liver, and so forth. If a surgeon used a healthy person undergoing routine tests as a resource for organs (thereby killing him, of course) all of the five would be saved.[46] Now if the pedestrian in the bridge scenario has sufficient reason to consent to being used as a train decelerator, then does not the healthy person have sufficient reason to be used as an organ resource? From Parfit's perspective, it would seem that the partial and impartial reasons that come into play in the cases are equivalent. The pedestrian and the healthy person both have strong partial reasons not to consent to being used; after all, to do so is for practical purposes to consent to being killed. But each also has enough impartial reason to consent, namely that five people will thereby be saved, such that overall he has sufficient reason to consent, Parfit seems committed to holding. So, assuming that, like the passerby, the surgeon is attempting to treat people only in ways to which they can rationally consent, she does not treat the healthy person merely as a means, even if before she succeeds in putting him under, he is begging for his life. According to *one* common sense understanding of treating others merely as means, the surgeon's treatment of the

---

[45] Parfit might respond that if we hold that view, then we are committed to thinking that in a traditional trolley case, the one killed does not have sufficient reason to consent to the points being switched. And that result would be implausible because then switching the points would be wrong, a result that many of us find unacceptable. But note that this response assumes that we endorse the Rational Consent Principle. It does not work if we do not endorse it. Second, there are crucial differences between the cases. One might say that we have impartial reason to refrain from consenting to being treated merely as a means and being killed that we do not have merely to being killed.

[46] For one source of this example, see Gilbert Harman, *The Nature of Morality* (Oxford: Oxford University Press, 1977), 3–4.

healthy person is a *paradigm example* of someone doing just that. And so, many of us think, is the passerby's treatment of the pedestrian on the bridge.

Parfit construes ordinary notions of treating others merely as means to be exhausted by a notion that takes the term "merely" in a literal way. On this notion, neither the passerby nor the surgeon treats the person she uses merely as a means, since she governs her use of him by the Consent Principle. She does not just treat the other as a means: she gives consideration to what the other has sufficient reason to consent to. Perhaps Parfit puts his finger on one ordinary notion of treating others merely as a means. But there is another notion, according to which "merely" is not to be taken literally. Put in general terms, to use someone merely as a means is, according to this notion, to use the person in a way that does not exhibit sufficient respect for the person's rational agency, that is, for her capacity to determine how she is used and to rationally pursue her ends. It is this notion that the Hybrid Account aims to capture.

In any case, Parfit offers a second ground for concluding that, contrary to what the Hybrid Account implies, the passerby need not be treating the pedestrian merely as a means. He asks us to imagine that the passerby has triggered the trapdoor, dropping the pedestrian onto the tracks. But if she could have done so, she would have then run in front of the pedestrian and used herself to slow down the train and save the five, rather than using him to do it. The Hybrid Account nevertheless classifies the passerby as using him merely as a means. But Parfit insists that she does not do this.[47] And that is because she is willing to die to prevent her (attempted) use of him from resulting in his death. But again, there is an ordinary sense in which we think the passerby does treat the pedestrian merely as a means. First, whatever else she would try to do if things were different, she does trigger the trapdoor, sending the pedestrian onto the track. That use of him exhibits the sort of lack of deference to others' rational agency that is a hallmark of treating them merely as means. Second, suppose in the transplant case that, if there were another surgeon available to do the operations, our surgeon would offer herself up as a resource for organs, and in so doing die, instead of using the healthy visitor for that purpose. Does that entail that when, as it turns out, she is the only surgeon available and carves up the visitor for his organs, she is not treating him merely as a means? Many of us, I venture, would reply no. We would say, rather, that if circumstances had been different, she would not have treated him merely as a means, but, as things stood, she did just that.

Although bridge cases do not really threaten the Hybrid Account, Parfit sketches others that might. For example, he describes a scenario along the following lines: After an earthquake, a father and his child are trapped in gradually collapsing wreckage, which will, as the father realizes, kill them unless they escape. The father cannot save his child's life except by using the body of a stranger, Black, as a shield, without her consent, in a way that would destroy one of her toes. If the father also caused Black to

---

[47] Parfit, *On What Matters*, 224.

lose another toe, he would save his own life.[48] (We can assume that Black's life is not itself threatened by the wreckage.) The father thinks that it would be wrong to save his own life by using Black as a shield and thereby crushing her toe. He believes that only the saving of a child's life could justify injuring someone else. So, without Black's consent, he goes ahead and saves his child's life by causing her to lose a toe.

Depending on how we fill in the details of the case, the Hybrid Account might imply that the father is treating Black merely as a means. First, let us specify that the father gives Black no opportunity to avert his using her as a shield. Black's not consenting to this use amounts to her ineffectually dissenting from it. Second, let us assume that it is reasonable for the father to believe that Black *cannot* share the father's end of her shielding his child from the falling debris. We can imagine that Black is yelling that, even in light of the child's plight, she is unwilling to give up her goal of getting out of the wreckage without significant injury, a goal that, as everyone involved is aware, she could not achieve if she joined the father in his pursuit of his end. If we make these improbable but realizable assumptions, then the Hybrid Account implies that the father is using Black merely as a means.

Is that conclusion implausible? It would be implausible to insist that the father *thinks of* Black solely as a tool, something that he would not wrong no matter how he used it. If he thought of Black in this way, then he would use her to save his own life, of course. However, we are asking about the status of the father's action. In using Black to shield his child, despite Black's protests, is the father using Black merely as a means? I believe that there is an ordinary sense in which the answer is yes. However, that does not entail that his action is wrong. For, as we have emphasized from the outset, using someone merely as a means is *pro tanto* wrong, but not necessarily wrong, all things considered. And in this case, intuitively speaking, it does not seem to be wrong all things considered. The lack of respect that the father exhibits towards Black's rational agency in treating her merely as a means seems to be outweighed in importance by the value that can be preserved (i.e., the child's life) only by his doing something that exhibits such a lack of respect.

Chapter 5 will investigate in more detail conditions under which treating others merely as means is not, all things considered, wrong. But we will arrive only at some broad guidelines, not an algorithm for deciding cases. It is worth remarking that Parfit himself seems to endorse a principle according to which "It is wrong to impose harm on someone as a means of achieving some aim, unless (1) there is no better way to achieve this aim, and (2) given the goodness of this aim, the harm we impose is not disproportionate, or too great."[49] But, as he acknowledges, this principle does not indicate which harms would be too great. "We would have to use our judgment here," he says.[50]

---

[48] ibid. 222.   [49] ibid. 229.   [50] ibid.

In this chapter we have been seeking a charitable specification of a Kantian notion of treating others merely as a means, namely one that allows us to claim plausibly that to treat another merely as a means is to *act (pro tanto) wrongly*. In his brief remarks in the *Groundwork* concerning how someone wrongly uses another with the help of a false promise, Kant hints at various ways in which we might understand conditions for treating another merely as a means. We might understand them in terms of the other's inability to share the agent's end in using him, or to avert the other's using him, or to rationally consent to it. Appealing to rational consent, at least how Parfit understands it, fails to yield a plausible specification of the Mere Means Principle. However, combining notions of possible end sharing and possible consent into the Hybrid Account does produce a sufficient condition for treating another merely as a means that both captures an ordinary understanding of it and has plausible implications.

4

Treating Consenting Adults Merely as Means

Our aim is to arrive at a charitable specification of the Mere Means Principle, according to which in treating another merely as a means, an agent is doing something that is (*pro tanto*) wrong. In Chapter 3, we developed a sufficient condition for an agent's using another merely as a means: the Hybrid Account. In this chapter, we try to deepen our understanding of the Mere Means Principle by formulating a sufficient condition for an agent's using another, but *not* merely as a means.

We have already found implausible one account of conditions under which an agent does not treat another merely as a means (3.6). According to that account, suggested by Derek Parfit, an agent does not do so if her use of another is governed by the Consent Principle. This chapter explores two alternatives. Following a suggestion by Robert Nozick, we might propose the Not-Merely-As-Means Actual Consent Account: when an agent uses another, she does *not* use him merely as a means if he has given his voluntary, informed consent to her using him.[1] We refer to this account as the Actual Consent Account[N], in order to distinguish it from an account, which we explore in Chapter 5, that uses a notion of actual consent to specify a sufficient condition for an agent's treating another merely as a means. The Actual Consent Account[N] seems to have the virtue of being simple and direct. A second alternative is to invoke a notion of possible consent that we examined in Chapter 3, according to which an agent can consent to being used only if he can avert this use by withholding his consent to it. The Not-Merely-As-Means Possible Consent Account (or the Possible Consent Account[N], for short) holds that when an agent uses another, she does *not* use him merely as a means if it is reasonable for her to believe that he can consent to her use of him.

This chapter explores the plausibility of the Actual Consent and Possible Consent Accounts[N]. It begins (4.1–4.2) by sketching the Actual Consent Account[N] in more detail, setting forth and briefly defending notions of *voluntary* and *informed* consent. The

---

[1] Robert Nozick, *Anarchy, State, and Utopia* (New York: Basic Books, 1974), 31. I do not wish to imply that Nozick actually embraces this view. His remarks are brief and leave it unclear whether he does. Some philosophers (in my view incorrectly) attribute this view to Kant himself. See, for example, Mark Cherry, *Kidney for Sale by Owner: Human Organs, Transplantation, and the Market* (Washington, D.C.: Georgetown University Press, 2005), 98.

chapter then considers difficulties, suggested by Onora O'Neill, that such an account faces (4.3). They turn out to be less serious than it might appear. Moreover, analogous difficulties apply to the Possible Consent Account[N]. So O'Neill's reflections leave us with no grounds for preferring one account to another.

However, the chapter argues that there are such grounds (4.4). The Actual Consent Account[N] suffers from shortcomings to which the Possible Consent Account[N] is immune. The account has the unwelcome implication that certain ineffective or unnecessary attempts an agent makes at coercing or deceiving another to serve as a means to her ends do not amount to her just using the other. The Actual Consent Account[N] thus fails to realize its promise of giving us a simple yet plausible way to capture jointly sufficient conditions for an agent's using another, but not merely as a means. The account can be altered so that it no longer has the unwelcome implication in question, but the altered account differs little, with respect to both complexity and content, from possible consent accounts.

There is an orthodox Kantian perspective from which even a modified actual consent account is inadequate. According to this view, an agent can treat another merely as a means and thereby act wrongly if she plans, under certain possible conditions, to preclude his voluntary, informed consent to her use of him, but she never actually attempts to preclude such consent. The chapter tries to arrive at a sympathetic understanding of this view, but in the end rejects it (4.5).

The remainder of the chapter (4.6) focuses mainly on pinpointing and resolving a difficulty faced by both actual and possible consent accounts. They both imply, implausibly, that an agent does not treat another merely as a means when she profits from vulnerability in the other that she (the agent) has induced.

As the chapter unfolds, it will be helpful to keep in mind two criteria for the success of accounts of treating others merely as means (Chapter 1). Accounts should, to a significant extent, square with the judgment of reflective common sense regarding circumstances in which a person uses (or does not use) another in this way. But it would be unreasonable to expect them to accord entirely with ordinary thinking. For one thing, it is not clear that there is any one notion for them to accord with: there might not be a *univocal* ordinary concept of "just using" another. Second, accounts of treating others merely as means should minimize appeal to concepts that are just as elusive as our notion(s) of treating others merely as means. If they rely on such concepts, they are unlikely to be illuminating.

4.1 Voluntary, Informed Consent

According to the Actual Consent Account[N], when an agent uses another, she does *not* use the other merely as a means if the other has given his voluntary, informed consent to her using him. An agent uses another just in case she intentionally does something to the other in order to realize her end, and she intends the presence or participation of (some aspect of) the other to contribute to the end's realization (3.1). For example, a tourist uses a police officer when she asks him for directions to get to the train station. But a policeman in pursuit of a suspect does not use a bystander he pushes out of his way; for he does not intend any aspect of the bystander to contribute to his reaching his goal.

For the sake of simplicity, let us assume throughout that the Actual Consent Account[N] is meant to apply only to competent adults. An agent is competent only if he has the capacity to decide, based on his own values as well as information available to him, whether to use others and whether to consent to their using him.[2] Of course, we need to get a reasonably clear idea of what *voluntary, informed* consent amounts to in the Actual Consent Account[N]. Notions of such consent have a central role in discussions of the ethics of medical treatment and research, as well as in rules governing these enterprises. But since the Actual Consent Account[N] is supposed to apply to all cases of one agent (i.e., competent adult) using another, the notion of voluntary, informed consent we employ here needs to apply to all such cases.

A person's consent or agreement to an agent's using him is informed if he understands the agent's purpose in using him, what she plans to do to him, and how it is likely to affect him, it seems reasonable to say. For example, a person's consent to serve as a means for an agent to get over a fence might be informed as a result of her telling the person that she would like a lift over it so she can get the ball that fell on the other side. She could presumably assume that the person has background knowledge of the stress on the body involved in giving someone a lift. But *how much* does a person need to know about the ends an agent plans to pursue in using him in order to give his informed consent to this use? Does he need to have knowledge only of the agent's proximate end or of her ultimate end as well? For example, does the person give his informed consent to an agent's using him to get over the fence if she informs him that she aims to retrieve the ball, but not also that she aims to retrieve it in order to get revenge on her neighbor by throwing it through his window? Let us set aside this issue here, since we discuss it in detail in the next section.

A defender of the Actual Consent Account[N] would presumably contend that a person's consent to another's use of him fails to be voluntary if it results from the agent's coercing her. The concept of coercion (assuming there is a univocal one) might be as difficult to specify with precision as that of treating others merely as means. But let us assume that the defender of the Actual Consent Account[N] adopts a simple account of coercion, according to which an agent's coercing another involves his threatening her, that is, claiming that, unless she does something, he will make her worse off than she would be without his involvement. According to the simple account, a victim does not give her voluntary consent to a mugger's demand for her money when he threatens to injure her unless she gives it up. In the next section, we will consider a more complex rival to the simple account. In any case, the defender of the Actual Consent

---

[2] For discussion of competence as it applies in healthcare settings, see Dan Brock, *Life and Death: Philosophical Essays in Biomedical Ethics* (Cambridge: Cambridge University Press, 1993), 36–43.

Account[N] might continue, the voluntariness of a person's consent to an agent's using him is not vitiated just by its being the case that the person desperately wants or needs the agent to do something for him and will be worse off if the agent does not do it.[3] For example, a motorist's being stranded on a sweltering desert highway does not render it impossible for him to agree voluntarily to a tow truck driver's picking him up and thereby using him to make a profit. Or suppose a tourist on vacation in a developing country has a heart attack and might die without emergency surgery. The tourist's situation does not preclude him from consenting voluntarily to a surgeon's performing a life-saving operation for her usual fee. Of course, there are other ways in which consent might fail to be voluntary, the defender might conclude. Illness, injury, or brainwashing might prevent someone from doing anything voluntarily, including consenting to be used by another.

A defender of the Actual Consent Account[N] might face the following objection to this view of voluntary consent. In the mugging case, the victim has a choice between losing his life and giving up some of his money. In the surgery case, the victim also faces a choice between these two outcomes. What basis do we have for agreeing with the defender's conclusion that what the victim does in the former case is involuntary but in the latter voluntary? The conclusion seems arbitrary.

Of course, if the defender altered his view and held that the consent of both the heart attack and mugging victims to be used for another's financial gain was voluntary, then he would no longer be vulnerable to the charge that he was making an arbitrary distinction between the two cases. But the defender would presumably not wish to do this. For, assuming that the mugger disclosed the nature of the use he was going to make of the victim (e.g., taking his wallet to get some cash), an implication of the Actual Consent Account[N] would then be that he was *not* treating the victim merely as a means. But that implication is simply not plausible. It is central to a common notion of treating others merely as means that the mugger is treating his victim in this way.

The defender might try to bolster his account against a charge of arbitrariness by maintaining that the action of neither victim was voluntary. But this response would also have problematic implications. First, the Actual Consent Account[N]'s force would be diminished. Suppose the surgeon tells the heart attack victim that she would operate for her usual fee, informs him of the procedure she plans and its likely effects on him, and proceeds only on condition that he agrees with all of this, which he ends up doing. The Actual Consent Account[N] would not have what many of us take to be the plausible implication that the surgeon is not treating the victim merely as a means. Second, suppose that the heart attack victim refuses the surgeon's offer of life-saving treatment for a fee. (We might imagine that he is old and thinks that the beautiful country he is visiting, which is the home of his ancestors, would be a good place to die.) How could the defender avoid the conclusion that his *dissent* fails to be voluntary? It

---

[3] Alan Wertheimer, *Exploitation* (Princeton, N.J.: Princeton University Press, 1996), 110.

seems odd to insist that a victim cannot voluntarily consent to a procedure that would save his life but can voluntarily consent to forgo such a procedure. Perhaps the defender would say that he can voluntarily consent to *neither*. But then, contrary to an entrenched view, a doctor would never derive any justification for refraining from giving a patient (e.g., someone with terminal cancer and at most weeks to live) a life-saving procedure (e.g., yet another painful operation) from the patient's voluntarily dissenting from receiving it. Finally, the defender would be forced to answer a difficult question: when *does* one's allowing an agent to use him count as voluntary? Assuming that one believes that some actions are voluntary, it would be wildly implausible to insist that one never voluntarily consents to another's using him. Is one able to voluntarily consent only when failure to agree to be used would not be too costly in terms of one's well-being? What, then, is too costly? Is a person able to give his voluntary consent to pay a surgeon to save his leg, but not to save his life?

Rather than embracing either the unappealing conclusion that both of our victims give their voluntary consent to being used or the equally uninviting one that neither do, the defender of the Actual Consent Account[N] should, I believe, insist that it is not arbitrary to say that the one (the heart attack victim) gives his voluntary consent, while the other (the mugging victim) does not. The cases differ from one another with respect to the victims' bases for consenting to be used. In one, the victim consents as a result of a mugger's claim that unless she gives him her money, he will make her worse off than she would be without interacting with him. In the other case, the victim's consent is based on no such threat. The voluntariness of a person's consent is not lost just because he is, like the heart attack victim, in a bad way—assuming, of course, that his condition is not so poor that he has lost his decision-making competence. But the voluntariness of a person's consent to an agent's use of him is lost if it is based on the agent's threat to make him worse off.

4.2 Coercion

Someone's consent to be used fails to be voluntary if it results from coercion, according to the Actual Consent Account[N]. A defender of this account would, we suggest, employ a simple notion of coercion. On this notion, coercion involves an agent's threatening another, that is, claiming that, unless she does something, he will make her worse off than she would be without his involvement. Franklin Miller and Alan Wertheimer reject this notion of coercion. In its place, they argue that we should adopt a "rights-violating" account, which states: "A coerces B to do X in a way that invalidates B's consent only if (1) A proposes or threatens to violate B's rights or not fulfill an obligation to B if B chooses not to do X and (2) B has no reasonable alternative but to accept A's proposal."[4] Miller and Wertheimer leave unspecified what B's rights

---

[4] Miller and Wertheimer, 390.

are vis-à-vis A as well as what obligations A has to B. The authors claim that the rights-violating account has two advantages over the simple account. First, the former but not the latter allows us to avoid the conclusion that when an agent has a right to give another a choice between two alternatives both of which will render him worse off, he is coercing the other and invalidating his consent. Second, the rights-violating account, but not the simple account, allows us to conclude that an agent is coercing another in certain cases in which she has an obligation to make the other better off, but will do so only on condition that the other benefit her in some way.

In my view, the advantages, if any, of the rights-violating account over the simple one are not significant. Moreover, the rights-violating account suffers from an important disadvantage relative to its rival. The rest of this section defends these views. This defense is important because later in the chapter we employ the notion of voluntariness (and the simple notion of coercion) introduced in the last section. But readers not interested in detailed discussion of coercion might skip to the next section.

Miller and Wertheimer illustrate the first alleged advantage of their account with a case of plea bargaining. Plea bargaining is "the process by which the defendant in a criminal case relinquishes the right to go to trial in exchange for a reduction in charge and/or sentence."[5] Suppose a prosecutor tells a defendant that he can either plead guilty to a charge and receive a lenient sentence or she will take him to trial. According to the simple account, the prosecutor is coercing the defendant; for she leaves him with two options, both of which are worse than his status quo. However, the rights-violating account does not have this implication, according to Miller and Wertheimer. In making her proposal, the prosecutor is not threatening to violate the defendant's rights, they imply. The alleged advantage that the rights-based account has over the simple account is that the rights-based account implies that the defendant might be giving valid consent if he agrees to plead guilty as a result of the prosecutor's proposal, whereas the simple account implies that his consent would be invalid. Miller and Wertheimer seem to think that we would reject the notion that his consent would be invalid. Invalid consent presumably fails to contribute to the moral legitimacy of what one consents to. For example, a victim's consent to give her kidnapper access to her bank account would not contribute any moral legitimacy to his use of her money. So, Miller and Wertheimer suggest, if we held the consent a defendant gives to a plea deal to be invalid, then we would have to agree, contrary to our considered views, that his consent would not contribute any moral legitimacy to his punishment.

In response, we have been employing a notion of voluntariness meant to be suitable for accounts of using another, but not merely as a means. On the notion we have been employing, an action does not count as voluntary if it results from coercion. If, as a result of a plea bargain as opposed, say, to a desire to acknowledge his misdeed and start making amends for it, a defendant confesses to a crime, then his confession is coerced,

[5] Candace McCoy, *Politics and Plea Bargaining* (Philadelphia: University of Pennsylvania Press, 1993), 50.

according to the simple account of coercion. So his confession is not voluntary, according to our notion of voluntariness. Miller and Wertheimer assume if consent is not voluntary, then it is not valid.[6] One might, however, maintain that in some contexts consent that is not voluntary, at least in the sense we have been employing, can nevertheless be legitimate. More concretely, a defendant who pleads guilty as a result of a plea bargain has been coerced and has thus not acted voluntarily in the sense of voluntariness we have embedded into our account of using others but not merely as a means. Nevertheless his guilty plea has been "free enough" to take as a basis for holding him accountable for a conviction on a lesser charge. Some might offer a consequentialist justification for holding coerced guilty pleas to be legitimate: providing a trial for every defendant who does not wish to plead guilty for reasons of conscience would be too expensive for the judicial system to handle.[7] The coercive element of plea bargaining is not ideal, but we have no realistic choice but to practice it, one might say.

An alternative to this response would be to say that since defendants who confess to crimes solely as a result of plea bargaining do *not* voluntarily plead guilty, their pleas carry no moral weight in justifying their punishment. This conclusion has some credibility. There are reasons to think that plea bargaining is a morally dubious practice.[8] In the United States, for example, a defendant facing a mandatory sentence of decades in prison if a trial leads to his conviction for drug trafficking might have a prosecutor assure him that if he pleads guilty to a lesser offense, he will serve only two to three years. Is it far-fetched to hold that the defendant's confession to the lesser offence fails to contribute to the moral justification of his ending up in prison? Some believe that despite the prevalence of plea bargaining in the United States' system of criminal justice, it ought to be eliminated. If we draw this conclusion, then the alleged advantage of the rights-based notion of coercion over the simple notion dissipates. This is not the place to enter into detailed debate regarding the morality of plea bargaining. But it does not seem to be a clear advantage of the rights-based account over the simple account that the former but not the latter is free from the implication that in confessing to a crime as a result of plea bargaining one is being coerced.

Miller and Wertheimer illustrate the second alleged advantage of the rights-based account with a case in which a doctor who works for a national health service is obligated to treat a patient free of charge. But the doctor proposes not to treat the patient unless he pays her. According to the simple account, the doctor is not coercing the patient; for she is not threatening to make him worse off than he presently is unless he pays her. However, assuming that the patient has no "reasonable alternative" but to pay the doctor (e.g., there is no other doctor available who would fulfill the obligation

[6] Miller and Wertheimer, 390.
[7] For discussion of such an argument, see Candace McCoy, "Plea Bargaining as Coercion: The Trial Penalty and Plea Bargaining Reform," *Criminal Law Quarterly* 50 (2005): 80.
[8] See McCoy, "Plea Bargaining as Coercion," 67–107.

to treat the patient for free), the rights-based account allows us to conclude that the doctor is coercing the patient, for the patient has a right to receive treatment from the doctor.

Perhaps this is a genuine advantage of the rights-based over the simple account. But I wonder how significant the advantage is. Granted, proponents of the simple account cannot say that the doctor is coercing the patient. Yet they are perfectly free to say that she is exploiting him, that is, taking unfair advantage of him and thereby acting wrongly. For some it might strain ordinary usage to refrain from saying that the doctor coerces the patient. But such strain does not seem to be a high price to pay for adopting the simple account which, as I now try to show, has a significant advantage over the rights-based account.

The advantage is that the simple account is more useful than the rights-based account. For it appeals to fewer unclear and disputed moral concepts than its rival. Granted, the simple account does invoke the notion of well-being. It states that an agent coerces another if she threatens to make her *worse off* if he does not do something. But the rights based account appeals to the notions of rights, of obligations, and of "reasonable" alternatives.

In appealing to this last notion the account is implicitly invoking a concept of well-being. Miller and Wertheimer mention two understandings of when someone lacks a reasonable alternative to do something. According to the first, "one has no reasonable alternative when the distance between the alternatives is so great that it would be irrational to reject the better alternative, and (virtually) no one would do so."[9] At least one of the reasons one alternative can be irrational to reject is that it contributes (or promises to contribute) so much more to a person's well-being than another alternative. To cite Miller and Wertheimer's own example, suppose someone offers you $1,000 to mow his lawn, which is a one-hour job. Your alternative of turning him down might not be a reasonable one for you to take—presumably because you could really use the money. Miller and Wertheimer's second understanding of lacking a reasonable alternative even more obviously implicitly invokes a notion of well-being. According to this understanding, someone has no reasonable alternative but to do something if not doing it would leave him in "dire straits." For example, someone might have no reasonable alternative but to undergo surgery to remove a cancerous tumor. So we can already see that, like the simple account of coercion, the rights-based account invokes a notion of well-being.

But the rights-based account also invokes two other disputed concepts in normative ethics, namely the notion of one person's having an obligation to another and, of course, the notion of violating another's rights. These two notions are at least as much in need of clarification as that of coercion itself. So it is hard to see how an account of (consent-invalidating) coercion that appeals to them can itself be very helpful. For

<hr />

[9] Miller and Wertheimer, 391.

example, say that a fisherman is out to sea, hurrying to bring in his crab traps, when he receives a distress call from a yacht with engine trouble. If he rescues the passengers, he and his crew will lose a third of a season's profits. If he does not rescue the passengers, they will, at the very least, suffer for several days from cold and hunger until another boat can reach them. Is the captain coercing the passengers on the yacht if he tells them that he will rescue them only if they reimburse him for his losses? Let us assume that the passengers have no reasonable alternative but to agree to reimburse him. If the captain has an obligation to rescue the passengers or if they have a right to be rescued, then the rights-based account would imply that he was coercing them; if he has no such obligation and they have no right to be rescued, then he was not coercing them, according to the rights-based account. But does the captain have this obligation or do the passengers have this right? The account leaves these questions unanswered. In effect, it leaves us having to invoke accounts of moral rights and obligations in order to determine whether the passengers have been coerced. In order to be illuminating, accounts of concepts in normative ethics need to minimize appeal to normative concepts. The simple account does a better job of this than the rights-based account. And in my view this advantage of the simple account outweighs the two ways in which, according to Miller and Wertheimer, the rights-based account is superior. So we will continue to take the Actual Consent Account^N to invoke a notion of voluntary consent that holds that an agent fails to give his voluntary consent to being used if her consent is coerced, according to the simple account of coercion.

4.3 Shortcomings of Actual Consent?

Onora O'Neill suggests reasons for rejecting the idea that in specifying the Mere Means Principle we should invoke a notion of actual consent. First, she implies that doing so would render the principle too unwieldy to use in practice. It is difficult to determine when individuals give their informed consent to being used in a particular way, she asserts.[10] Even if they are told in detail what will be done to them and to what purpose, have they really understood? In some cases it does indeed seem difficult to know whether they have.

But this point is not sufficient to undermine the Actual Consent Account^N. First, that in some cases we are unable to determine with confidence whether an agent has given his informed consent to be used in a certain way is consistent with our being able to do so in the vast majority of others. O'Neill gives us no reason for *general* skepticism regarding our ability to know whether someone has understood a proposal to use him. Second, even if it were *always* challenging to determine whether someone had given his voluntary, informed consent to being treated as a means, that would not entail it to

---

[10] Onora O'Neill, *Constructions of Reason* (Cambridge: Cambridge University Press, 1989), 108.

be mistaken to hold that no one who did in fact give such consent is being used merely as a means.

Third, the same sort of issue O'Neill points to regarding actual consent arises in the context of possible consent. Whether an individual really has an opportunity to avert someone's treatment of him as a means is often a function of whether he grasps what the proposed treatment is. But it can be difficult to discern whether he grasps this. Does the financial advisor's client understand the myriad ways she uses him to make a profit well enough to prevent it? O'Neill says that patients "cannot easily understand complex medical procedures; yet if they consent only to a simplified account, they may not consent to the treatment proposed."[11] Indeed, it can be challenging to ensure that patients give informed consent to complex medical procedures and, of course, challenging to determine whether they have done so. But analogous points apply to the Possible Consent Account[N]. If a doctor gives a patient only a simplified description of a complex medical procedure, then the patient might not have an opportunity to avert the doctor's performing *that procedure* by withholding his agreement to it. His having such an opportunity depends in part on his understanding what the procedure is. And it can be difficult to ascertain whether he understands this. In the context of medical treatment and research, empirical investigators are developing methods for measuring both the extent of patients' understanding of interventions to which they consent and how this understanding varies in relation to various methods medical professionals use to inform them regarding the interventions. Data collected from these studies has helped investigators to construct materials that promote patient understanding.[12]

O'Neill suggests that there is a "deeper" problem with the Actual Consent Account[N]. "When we consent to another's proposals," she says, "we consent . . . only to some specific formulation of what the other has it in mind to do. We may remain ignorant of further, perhaps equally pertinent, accounts of what is proposed, including some to which we would not consent."[13] If we would not consent to some proposed treatment of us as a means under these further descriptions, would our consent to the treatment under the description we are familiar with suffice to make it such that we are not being used merely as a means? Let us call this question the description concern. Note that the description concern does *not* stem from one or the other description of the proposal containing false information.[14]

---

[11] O'Neill, *Constructions of Reason*, 108.

[12] See Gerd Gigerenzer et al., "Helping Doctors and Patients Make Sense of Health Statistics," *Psychological Science in the Public Interest* 8 (2008) and Philip Candilis and Charles Lidz, "Advances in Informed Consent Research," in *The Ethics of Consent*, ed. Franklin Miller and Alan Wertheimer (Oxford: Oxford University Press, 2010).

[13] O'Neill, *Constructions of Reason*, 108.

[14] Of course, if we agree to be treated as a means after having been presented with false information, it makes sense to doubt whether our agreement has really been informed. And if our agreement has not been informed, then the Actual Consent Account[N] would be free of the implication that we have not been treated merely as means.

The description concern might come about as a result of framing effects.[15] People who receive logically equivalent information regarding the outcome of taking an option choose differently depending on how that information is presented to them, according to many studies. For example, in one well-known study, populations of patients, physicians, and graduate students were asked to indicate whether they would opt for surgery or radiation as a treatment for lung cancer. In each population, experimental subjects were less likely to indicate that they would take the option of surgery if they were told that there was a 10 per cent chance of dying from it than if they were told that there was a 90 per cent chance of surviving it.[16] Other studies have generated similar results.[17]

What are we to make of such results in the context of developing sufficient conditions for an agent's using another, but not merely as a means? Suppose that an orthopedic surgeon informs a patient about an operation she proposes to do on him. She explains the procedure, the improvement in his mobility she hopes to accomplish with it, and so forth. She also informs him of the operation's risks, including that there is a 99 per cent chance of his surviving it. The patient gives his voluntary consent to go ahead with it, allowing the surgeon to treat her as a means to financial gain. Let us assume that we (somehow) know the following: the patient would *not* have proceeded with the operation had the surgeon informed him that there is a 1 per cent chance of his dying from it. Even in light of such knowledge, we could not say that the patient's consent to the operation was uniformed. The surgeon did not falsify the risks of the surgery. If the surgeon was unaware of framing effects, including, of course, of their salience in this case, did she treat the patient merely as a means, according to our reflective common sense? The answer is clearly no, I believe. Moreover, unless the surgeon's ignorance of framing effects amounted to negligence, it is hard to see how she did anything morally wrong.

But now consider a second scenario, that is, one just like the first except that the surgeon is aware of framing effects and, in order to increase the likelihood that the patient will agree to the operation and thereby contribute to her bottom line, she describes its risks in terms of a 99 per cent survival rate rather than in terms of a 1 per cent mortality rate. It would be a stretch to insist that the patient failed to give his informed

---

[15] O'Neill does not explicitly mention framing effects in this context.

[16] See Barbara McNeil et al., "On the Elicitation of Preferences for Alternative Therapies," *New England Journal of Medicine*, 306 (1982). The patients had not been diagnosed with lung cancer, but rather with various chronic conditions.

[17] See, for example, Katrina Armstrong et al., "Effect of Framing as Gain versus Loss on Understanding and Hypothetical Treatment Choices: Survival and Mortality Curves," *Medical Decision Making* 22 (2002) and Sammy Almashat et al., "Framing Effect Debiasing in Medical Decision Making," *Patient Education and Counseling*, 71 (2007). Researchers have isolated phenomena that bear a family resemblance to framing effects. For example, in a recent study parents expressed significantly greater willingness to enroll their children in a medical experiment if it was described as a "research study" than if it was described as a "medical experiment." See Stephen Cico, Eva Vogeley, and William Doyle, "Informed Consent Language and Parents' Willingness to Enroll Their Children in Research," *IRB: Ethics and Human Research* 33 (2011).

consent. But has the surgeon treated him merely as a means, intuitively speaking? If we were to conclude that she has, then the viability of the Actual Consent Account[N] would be threatened; for it implies that the surgeon has not done so. But the Possible Consent Account[N] would have the same implication, and its viability would also be threatened. The patient's being more likely to agree to the operation when the surgeon frames its risk in positive terms does not prevent him from averting the surgeon's use of him simply by withholding his consent from it. In any case, I believe that most of us would answer the question negatively: the surgeon is not treating the patient merely as a means. She might be (wrongfully) manipulating or exploiting him, but it seems to be overly dramatic to characterize her as *just using* him.[18] It is, of course, consistent with denying that the physician is treating the patient merely as a means to hold that physicians (and medical researchers) have a duty to present information in ways that minimize framing effects.[19]

A more pressing version of the description concern might arise when a person's consent to being treated as a means by an agent is contingent on the agent's aims being described in less rather than more detail, even though both ways of describing the aims are factually correct. We alluded to this issue above. A person might agree to a proposal to help someone get over a fence to retrieve a ball. But suppose that the proposal might also be described differently, namely as one to help someone get over the fence to retrieve a ball so that he can throw it through the neighbor's (closed) window. If a person agrees to the proposal under the first description, but would not agree to it under the second, we might conclude that he is being treated *merely* as a means, even though he has consented to being treated as a means.

If it seems not at all plausible to draw this conclusion in this case, imagine a conservationist who is trying to preserve tiger habitat from being taken over by nearby villagers. She enlists one of the village elders to act as a liaison. She would like his help, she tells him, in order to pinpoint what the villagers really need. He gives her information and she provides well-targeted aid. However, she does not tell him that her ultimate end is to preserve tigers and that she takes as a means to this end making the villagers prosperous enough not to need to appropriate more jungle.[20] If the conservationist had proposed to the elder to use him to attain this ultimate end, he might not

---

[18] Those who do believe that, in the second scenario, the surgeon is treating her patient merely as a means will, as I have suggested, question the plausibility of the Actual Consent and Possible Consent Accounts[N]. But the accounts might be modified to address their concerns. For example, the Actual Consent Account[N] might be revised into the following: When an agent uses another, she does *not* use him merely as a means if he has given his voluntary, informed consent to her using him and, in the process of getting the consent, the agent has taken reasonable steps to minimize the influence of framing effects and other cognitive biases. The Possible Consent Account[N] could, of course, be modified in an analogous way.

[19] A growing literature describes methods of minimizing them. See, for example, Racio Garcia-Retamero and Mirta Galesic, "How to Reduce the Effect of Framing on Messages About Health," *Journal of General Internal Medicine* 25 (2010) as well as Almashat et al., "Framing Effect Debiasing in Medical Decision Making."

[20] This sort of example is not fanciful as an interview with Jane Goodall reveals (Gary Weitzman, "Dr. Jane Goodall," in *The Animal House* (2009)).

have consented to this use; for he might despise tigers on account of their deadly attacks on his people. If the elder would indeed have refrained from consenting, some of us might, intuitively speaking, conclude that the conservationist treated him merely as a means, even though he consented to be used as a source of information.

A defender of the Actual Consent Account[N] might insist that the elder did not really give his *informed* consent. But then how much information does one need in order to be informed? Does one need to know not only the agent's proximate end, but his further and ultimate ends as well? The danger here is that we set the bar for being informed so high that a person rarely counts as being informed.

A different defense might be to alter the Actual Consent Account[N] to maintain the following: An agent who uses another does not use her merely as a means if it is reasonable for her to believe that the other gives his voluntary, informed consent to her use of him. It is reasonable for an agent to believe that another gives his informed consent only if it is reasonable for her to believe both that the other is aware of her proximate end and that if he was aware of her further ends, he would still agree to the agent's using him. Here "reasonable" is being employed in our non-moral sense, according to which whether it is reasonable for an agent to believe something depends on his epistemic situation. On this account, it might not be reasonable for the conservationist to believe that the village elder gives his voluntary, informed consent to her using him for information. And it might not be reasonable for the future window breaker to believe that the other gives his voluntary, informed consent to his using him to get over the fence.

This revision would come at a cost. In some cases, the account would fail to yield the verdict that an agent had not treated another merely as a means, even though many of us would embrace this verdict. Suppose, for example, that a marketing executive is using consumers to further her company's ends. She sends out discount coupons for a product in an effort to get consumers to try it, form a habit of purchasing it, and drive a particular competitor out of business, thereby dramatically increasing her profits. Since the competitor is an old, family run company for which many have warm feelings, it is reasonable for the executive to believe that some consumers would not take advantage of the coupon if they were aware of her end of vanquishing the competitor. Nevertheless, many of us believe that in sending out coupons to get consumers to purchase her product, the executive is not treating them merely as means. But the revised Actual Consent Account[N] would fail to imply that she does not do so. Sometimes an agent's reasonable belief that someone she is using would not agree to be used if he was aware of some of her further ends seems compatible with her nevertheless not using this person merely as a means. But the revised account we are considering fails to register this compatibility.

The revised account was designed to respond to a difficulty posed for the Actual Consent Account[N] by cases like those of window breaking and tiger saving described above. But it is open to proponents of actual consent to respond that the problem applies just as much to possible consent as it does to actual consent accounts. Suppose that an agent has had an opportunity to consent to being used as a prop for someone to get over a fence to retrieve a ball. That does not mean that she has had an opportunity to consent to being used as a prop for someone to retrieve a ball so that she can break the neighbor's window with it. She might not have known that window breaking was in the cards. In that case, when she is used to get over the fence, is she treated merely as a means?

If one has the intuition that the answer is yes, then one might make a modification to the Possible Consent Account[N] analogous to the modification we suggested to the Actual Consent Account[N]. One might say, for example, that an agent who uses another does not use her merely as a means if it is reasonable for her to believe that the other can consent to her use of him. Moreover, it is reasonable for an agent to believe this only if it is reasonable for her to believe both that the other is aware of her proximate end, and that if he was aware of her further ends, he would not have chosen to avert the agent's use of him.

For the sake of simplicity, let us assume in this chapter that in order to count as giving his informed consent to be used in a certain way, a person need have knowledge of the agent's proximate end in using him, but not of his further ends. This assumption makes no substantive difference to the discussion that follows.

In sum, O'Neill points to some difficulties with formulating and applying actual consent accounts. But since analogous difficulties apply with equal force to possible consent accounts, the difficulties do not give us good reason to prefer one approach over the other.[21] The description concern to which O'Neill alludes is worthy of attention. But when it is triggered by framing effects, it seems not to undermine the Actual Consent Account[N]. When the concern stems from variously articulated descriptions of ends, it seems more pressing. Yet both the Actual and Possible Consent Accounts[N] can be modified to accommodate it, albeit at a cost. In the end, O'Neill herself has apparently come to believe that the description concern is not too serious in practice. In the context of medical treatment and research, she says that consent requirements—and they are *actual* consent requirements—function "reliably as an everyday way of permitting action that would otherwise violate important norms and standards."[22] Actions such as giving people dangerous chemicals to ingest or cutting them with sharp instruments would, without their informed consent, presumably

---

[21] Both actual and possible consent accounts of when an agent uses another but not merely as a means imply that there are circumstances in which an agent might kill another for his body parts without "just using" the other. We can, for example, imagine that a competent adult gives his informed, voluntary consent to be killed by a transplant surgeon so she can use his organs to save the lives of five other people, and that the surgeon gives him ample opportunity to avert his use of him for this purpose. Some, for example, Peter Schaber ("Using People Merely as a Means") hold that, if the surgeon goes ahead with the operation in order to save the five, she treats the person merely as a means. But while we might conclude that the surgeon is acting wrongly, I do not believe we would say that she is treating him merely as a means. The surgeon might be doing that, for example, if she did not give him the opportunity to prevent her from harvesting his organs.

[22] Manson and O'Neill, *Rethinking Informed Consent in Bioethics*, 76, italics omitted. See also 96 and 149.

violate norms that prohibit recklessly endangering them, assaulting them, or, of course, treating them merely as means.

4.4 From Actual toward Possible Consent

Nevertheless, there is reason to move away from actual consent and toward possible consent in an account of conditions under which an agent uses another, but not merely as a means. Consider the case of a 65-year-old salesperson who believes that her company is trying to get her to retire by keeping its latest sales leads from her. Desperate to make a sale, she intends to use the office manager to get the leads. They are kept in a database, to which he has the password. She tells him that she really needs to close some deals, and unless he gets the leads for her, she is going to reveal to everyone in the office that he is gay. She believes reasonably, given her limited understanding of him and of the attitudes of her other co-workers, that this revelation would be damaging to his reputation. But the office manager takes the salesperson, whom he thinks of as a friendly colleague, to be making a misguided joke. Everyone in the office who cares to know is already aware of his sexual preference. And, he believes, she knows full well that it is company policy that all salespeople are to be granted access to the latest leads upon request. He gives her a puzzled look and agrees to get her the leads right away.

The salesperson receives from the office manager his voluntary, informed consent to her use of him to obtain the leads. He understands that she intends to use him for this purpose. Although she threatens to make him worse off if he does not give her the leads, it is not the threat, which he does not even perceive as such, that prompts him to agree to her use of him. He does so voluntarily. Yet, despite getting his consent for her use of him, many of us believe that the salesperson treats the office manager merely as a means and acts wrongly.[23]

It would be easy to multiply cases like this in order to illustrate that a person's giving his voluntary, informed consent to being used is compatible with his being used merely as a means. The cases would include ones in which an agent attempts to coerce or deceive another into allowing her to use him, but in which the other allows himself to be used by her for reasons that have nothing to do with this attempt.

One might object to the notion that such cases illustrate this point. Inspired by Thomas Scanlon, one might claim, for example, that the salesperson does not *act* wrongly, that is, that what she does is not morally impermissible. After all, her coercive

---

[23] The Hybrid Account (3.5) might entail that she is doing just that. For it would presumably be reasonable for her to believe both that the office manager cannot consent to her use of him and that he cannot share her proximate end in using him. Regarding the former point, it would presumably be reasonable for her to believe that he cannot avert her use of him simply by withholding his consent to it, as opposed to withholding his consent and accepting a reduction *by her* of his well-being. Regarding the latter point, it might be reasonable for the salesperson to believe that in willing her to get the leads, the office manager would, in effect, be willing it to be impossible for him to accomplish an important end which, in this context, he would be unwilling to give up, namely, that of keeping his job.

threat is impotent. Nevertheless, her use of the office manager does manifest a moral fault. In light of her belief in the force of her threat, she ought to, but fails to, *see* the way she uses the manager as morally impermissible.

In order to assess this objection, let us return to its source. Scanlon introduces a distinction between factors that make an action wrong and factors that make it an action that the agent should see as wrong:

> This distinction is clearest in cases in which the action an agent proposes to take is in fact utterly harmless. A person who believes in voodoo, for example, may think that by sticking pins in a doll he is bringing about the agonizing death of his former girlfriend's new lover. But there is no reason to think that sticking pins into a doll is in fact harmful. So how could it be impermissible? There does seem to be something wrong with the action I have described. What is wrong with it, however, is not that it is impermissible but rather that the agent should (given his beliefs) see it as impermissible.[24]

Let us call the person practicing voodoo Kulev (which apparently means "harmless snake" in voodoo terminology). Scanlon seems correct to hold that Kulev should, morally speaking, see his action as morally impermissible. But, at least to many of us, he seems to be on shaky ground in suggesting that Kulev's action is actually morally permissible. After all, he is trying to bring about the agonizing death of an innocent person.

Scanlon supports his suggestion that Kulev's action is morally permissible with the assertion that there is "no reason to think that sticking pins into a doll is in fact harmful." But this assertion prompts the question: no reason *for whom* to think that? Granted, there is no reason for someone with knowledge of the context, including knowledge of the actual effects of Kulev's sticking pins into this wax figure, to believe that his doing so will kill the new lover. But that no person with such knowledge would believe his action to be harmful is compatible with its being wrong.

In May 2009, James Cromitie rigged two vehicles parked in front of a Bronx synagogue with powerful bombs, or so he believed. But, as someone with knowledge of the context would realize, the bombs were actually fake: Cromitie had purchased them from an FBI informant.[25] That there was no reason for a person with knowledge of the context to think that the devices were harmful does not entail that it was morally permissible for Cromitie to use them in an effort to maim and kill innocent people. There was presumably reason for Cromitie himself to believe that his manipulating the devices would harm others. Given his epistemic situation, it was reasonable for him to believe that they would explode.

In a similar way, given Kulev's epistemic situation, which is a function of his native abilities, his upbringing, and his circumstances, perhaps *he* has reason to believe that he

---

[24] Thomas Scanlon, *Moral Dimensions: Permissibility, Meaning, Blame* (Cambridge, Mass.: Belknap Press of Harvard University Press, 2008), 46.

[25] S. Sataline, C. Bray, and G. Fields, "Four Men Held in U.S. Terror Case," *The Wall Street Journal*, May 22, 2009.

can kill his ex-girlfriend's new lover by sticking pins in a wax representation of him. The strangeness of this belief might prompt the reaction that Kulev simply cannot have reason to embrace it. But is this belief really any less strange, from a scientific perspective, than the belief, apparently held by almost eight-in-ten Americans, that "miracles still occur today as in ancient times"?[26] In light of their upbringing, experience, and so forth, might not some of these Americans have reason to believe this? (Of course, an affirmative answer to this question in no way implies that we ought to embrace the belief as true.)

Scanlon's claim that Kulev does not act wrongly has some initial plausibility. But I suspect that for many this plausibility will wane after we uncover some unacknowledged views that influence assessment of the case. First, reflecting on the fact that *she herself* has no reason to believe that Kulev's stabbing a wax figure will result in the new lover's death, a person might fall into the view that *no one* has reason to believe this. But if no one has reason to believe this, the person might think, then Kulev's action is that of someone out of touch with reality, and thus an action not correctly categorized as wrong. The person might judge behavior to be morally permissible or impermissible only if it stems from a competent choice by an agent. If I am correct, however, Kulev might have reason to believe in the efficacy of voodoo. Second, someone might hold that if Kulev is mentally competent and is trying to kill an innocent person, then he is acting wrongly. But despite Scanlon's description, according to which Kulev believes that sticking pins in the wax figure will result in the death of his ex-girlfriend's lover, this person might find this belief so bizarre that she does not think Kulev really has it, and so does not think that he is really trying to kill anyone. If such a person embraces the possibility that Kulev genuinely has this belief—and I, of course, think he might— then she will judge his action to be wrong.

Let us now return to the case of the salesperson. Someone with knowledge of the context in which she tries to get the latest leads from the office manager would have no reason to believe her threat to have any causal power. Nevertheless, the salesperson has reason to believe that it does. Kulev and Cromitie's having reason to believe in the efficaciousness of their attempts to bring about harm supports the conclusion that the attempts are themselves wrong, not merely actions that each ought to see as wrong. In a similar vein, the salesperson's having reason to believe in the efficaciousness of her threat supports the notion that the way she uses the office manager *is* wrong, not merely something she should see as such. So the Scanlon-inspired objection does not invalidate the claim that an agent can use another merely as means (and thereby act wrongly) even if the other gives his voluntary, informed consent to being used by her. The Actual Consent Account[N] is vulnerable to a serious objection.

Of course, the salesperson case differs from the two others mentioned. We can imagine that Kulev is trying to use the demise of the new boyfriend to hurt his ex-

[26] "U.S. Religious Landscape Survey: Religious Beliefs and Practices: Diverse and Politically Relevant," (Washington D.C., 2008), 11.

girlfriend and that Cromitie is trying to use the killing and maiming of Bronx Jews to promote his religious ideology. In those cases, the envisaged use of a person would itself be harmful to the person. But using the office manager as a means to get leads would presumably not itself be harmful to the manager. What, in the misguided view of the salesperson, would harm him is the revelation of his sexual preference. This disanalogy between the cases would be important if the notion of using others merely as means included by its very concept only the harmful using of them. (There would then be little reason to believe that the salesperson was treating the office manager merely as a means.) But there is no such conceptual connection. Whether an agent treats another merely as a means is not a function of whether her use of the other actually harms him or is intended to harm him, but of *how* she uses him, which includes what she does or tries to do in order to make this usage possible. What the salesperson does in an effort to be able to use the office manager as a means makes it particularly plausible to conclude that she treats him merely as a means.[27]

In light of cases such as that of the salesperson, one might, of course, amend the Actual Consent Account[N]. Let us call the following account the Modified Actual Consent Account[N]:

An agent uses another, but does not use him merely as a means, if it is reasonable for her to believe that

the other gives his voluntary, informed consent to her using him, and

she has made no attempt to get him to agree to her using him other than that of trying to gain his voluntary, informed consent to her using him.

---

[27] Someone might object that the salesperson does not treat the office manager merely as a means, but rather tries but fails to do what would amount to treating her in this way. She would genuinely treat him merely as a means only if he had taken her threat to reveal his sexual preference seriously, according to the objection. I fail to see an intuitive basis for rejecting the idea that the salesperson treats the office manager merely as a means. Assuming that one uses another, it seems plausible to hold that whether one treats the other merely as a means can depend in part on characteristics of one's efforts to make the usage possible, regardless of whether the efforts do indeed make it possible. So, for example, assume that an auto mechanic uses a customer to make money. It seems plausible to hold that she can count as treating the customer merely as a means if she lies to him about what is wrong with his car in order to get him to authorize the repair, regardless of whether the lie in fact contributes to his authorizing the repair. (We can imagine that he authorizes the repair not at all as a result of what she tells him, but rather solely on the basis of his supposed insight into the workings of engines.) Whether or not the mechanic's lie leads to the customer being deceived, she *treats* the customer merely as a means, one might plausibly claim. If we reflect on it, many of us hold that some cases of treating another merely as a means bear a structural resemblance to lying. One can count as lying to another regardless of whether one succeeds in getting another to embrace a false belief. Similarly, one can count as treating another merely as a means regardless of whether one's attempt to make the use of the other possible really contributes to the realization of that use. Finally, focusing on the salesperson example, suppose that, contrary to my view, we can conclude from it that the salesperson tries to do what would amount to treating the office manager merely as a means, but not that she treats him merely as a means. The example would then not strictly speaking illustrate that the Actual Consent Account[N] is false. But assuming we agree that what the salesperson does is wrong, it would illustrate that a person can be acting wrongly by virtue of the way she uses someone, even if the person gives his voluntary, informed consent to this use. And that result seems significant in itself.

This account would avoid the implausible implication that the salesperson does not treat the office manager merely as a means. She has threatened him in order to get him to yield to her using him.

Notice how far we have come. We started with the notion that an agent does not use another merely as a means if the other gives his voluntary, informed consent to her using him. We started, we might say, with a *patient-focused* account. According to the Actual Consent Account^N, whether an agent treated another merely as a means depended primarily on the state of the other (the patient), that is, on whether he understood the use he was going to be put to and voluntarily consented to being put to that use. But we have found that, intuitively speaking, a patient-focused actual consent account is inadequate. Whether an agent uses another merely as a means can be a function not only of the other's state, but also of what the agent tries to do, regardless of whether it affects the other's state. What an agent attempts to do to another in order to make practically possible her use of him can influence whether she treats the other merely as a means.

Our modifications to the Actual Consent Account^N have, for practical purposes, rendered it hard to distinguish from the Possible Consent Account^N. An agent's attempt to get another to agree to her using him other than by trying to gain his voluntary, informed consent would likely amount to an attempt to deprive another of an opportunity to avert her use of him by withholding his consent to it. For example, the salesperson tries to get the office manager to give her the leads with the help of a threat, thereby trying to deprive him of the opportunity to prevent her use of him simply by dissenting from it. Conversely, central cases of an agent's preventing another from having an opportunity to consent to her use of him, that is, cases of deceit and coercion, involve attempts by the agent to get another to submit to her using him other than by gaining his voluntary, informed consent. If the agent successfully employs deception, then the other's consent is not informed. If she successfully employs coercion, then the other's consent is not voluntary. A defensible version of an actual consent account veers close to a possible consent account.

4.5 Orthodox Kantianism and Actual Consent

There is an orthodox Kantian perspective from which even the Modified Actual Consent Account^N is inadequate. According to this perspective, an agent can treat another merely as a means even if the other gives his informed, voluntary consent to her use of him and she never actually attempts to preclude him from giving such consent. A variation on our salesperson example will help to illustrate this possibility. Imagine that the salesperson suspects, but is far from sure, that the office manager will refuse to give her the latest leads. She acts on the following plan: she will ask him for the leads, and if he does not give them to her immediately, she will threaten to reveal his sexual preference. When she asks him for the leads, he gives them to her right away. So in the end she makes no threat. According even to the Modified Actual Consent

Account[N], it turns out that she is not using him merely as a means. Not only does he give his voluntary, informed consent to her use of him, but she makes no attempt to get him to agree to her usage of him through other means.

Those who adopt this orthodox Kantian perspective might champion a possible consent account of conditions under which an agent *does treat* another merely as a means that entails that the salesperson, in the latest variation of the example, is doing just that. (The Hybrid Account presented in Chapter 3 does *not* entail this; for it is reasonable for the salesperson to believe that the office manager can share the end she is pursuing in using him.) According to these Kantians, an agent does not use another merely as a means unless she uses him, of course. But the action of using itself, defined roughly as doing something to another in order to attain one's end, does not exhaust what is relevant to consider in deciding whether one is using another merely as a means. Also relevant is *how* one uses the other, including, according to them, which options one *intends* to leave for the other as one employs him as a means.[28]

But a critic might here insist: the salesperson would have treated the manager merely as a means if, after he had denied her request for the code, she had tried to force it out of him. However, since he voluntarily accepted her request, she did not treat him merely as a means. Her action regarding him was merely to ask him for the password in order to gain access to the database. And that surely doesn't amount to "just using" him.

Perhaps the critic is at least tacitly taking an impoverished view of the sort of act-descriptions we may plausibly employ in evaluation of moral permissibility. According to Kant, all actions, or, more precisely, all actions subject to moral evaluation, are done on maxims: self-given rules for acting. When an agent acts, he does so on a maxim in the sense that the maxim helps to generate the action. A maxim is not simply a description applied to an action after it occurs. On my interpretation, a fully-articulated maxim would describe an incentive for doing something in certain circumstances in order to attain some end, for example, "From self-love, I exercise during my free time in order to stay in shape." Kant's official view is that whether an agent's action is morally permissible depends on its maxim.[29] It is thus not surprising that the canonical formulation of the categorical imperative commands that we act only *on maxims* such that we can, at the same time, will that they become universal laws.[30]

Suppose that the salesperson is acting on a maxim, namely: "From self-love, given that I'm in financial trouble, I'll do anything to the office manager I need to do in order to secure a sale." Regardless of whether she actually threatens or harms the manager, in acting on this maxim she would, according to the Kantian view, be treating him merely as a means and thus acting wrongly. The principle that helps to generate her

---

[28] Of course, based on what she does in this example, the salesperson would in all likelihood not be legally vulnerable to a charge that she assaulted, coerced, or harassed the manager. But at issue here is moral, not legal permissibility.

[29] For justification of the interpretation given here of maxims and acting on them, see Samuel Kerstein, *Kant's Search for the Supreme Principle of Morality* (Cambridge: Cambridge University Press, 2002), 16–20.

[30] GMS 421.

action is that of doing anything she needs to do to the manager in order to attain her end. She thus uses him as one would use a tool—for example, a computer that one was willing to take apart, damage, and even destroy in order to recover data it contained. According to the Kantian view, the principle on which someone acts is constitutive for purposes of evaluation of moral permissibility of the kind of action she performs. If the salesperson had instead acted on the maxim "From self-love, given that I'm in financial trouble, I'll try my best, within the bounds of decency and legality, to secure a sale," then on this view she would have performed a different action.

Granted, from the perspective of a third party, the two actions would be indistinguishable. But that sort of indistinguishability fails to point to moral indistinguishability. From the perspective of a third party, a surgeon's making an incision in order to save someone's life might be indistinguishable from her making an incision in order to kill him. Yet, for the purposes of moral evaluation, the actions are obviously different.

Kant's account of acting on maxims faces serious difficulties. For example, when an agent acts, he does not always have in view a particular maxim. He nevertheless acts on one, according to Kant. But it seems that various different principles might have played a role in the generation of his action. When he told the homeless mother that he would not give her money, did he act on a maxim of discouraging begging for the sake of public order or rather on a maxim of maximizing his own happiness even if that means not helping others in need? How is the agent to know which one he acted on? And if he does not (or even cannot) know, then how can he assess the moral permissibility of what he did? Answering questions such as these poses a serious challenge for orthodox Kantian ethics.[31] In light of such challenges, it seems reasonable to doubt the contentions that every morally evaluable action is done on a maxim and every such action's moral permissibility must be assessed by appeal to its maxim.

But in order for the orthodox Kantians' main point to stand it is not necessary to accept these very general propositions. One must, however, embrace two claims. First, an agent can act on a maxim—in particular, in trying to get the code from the manager, a salesperson might act on the maxim described above. Second, her doing so would be relevant to evaluating the permissibility of her action; in particular, if she acts on this maxim she treats the manager merely as a means and thereby acts wrongly.

Of course, it is open to the critic of the Kantian view to insist that when in our example the salesperson acts on the maxim in question in trying to get the code, her action is morally permissible. It is merely her attitude, which is manifest in the principle she has adopted, that is morally problematic. Derek Parfit might make this objection. Sketching an example we have already encountered, he writes:

> Consider some gangster who . . . regards most other people as a mere means, and who would injure them whenever that would benefit him. When this man buys a cup of coffee,

---

[31] For discussion of further problems with maxims, see Rüdiger Bittner, *Doing Things for Reasons* (Oxford: Oxford University Press, 2001), 43–8.

he treats the coffee seller just as he would treat a vending machine. He would steal from the coffee seller if that was worth the trouble, just as he would smash the machine . . . [W]hat is wrong is only his attitude to this person. In buying his cup of coffee, he does not act wrongly.[32]

Parfit's gangster uses the seller to get coffee. Moreover, he has a certain disposition with respect to her. He "would" steal from her "if that was worth the trouble." Parfit's contention that the gangster, as he describes him, does nothing morally impermissible in buying his coffee strikes me as very plausible.

But perhaps there is a salient difference between the gangster and the salesperson. The salesperson acts on a specific plan to force the office manager into submitting to her use of him, while the gangster acts on no specific plan. Suppose that the gangster walks into the coffee shop and thinks to himself: "I really want a double mocha latte, but I don't have enough cash for it. If this barista won't give me the coffee for what I do have, I will force her, with physical violence if necessary, to give it to me." He then gives his order to the barista and tells her that he's a bit short. But she responds that that's no problem at all and gives him his drink. Here it is the case not only that the gangster would steal from the barista if she did not allow him to underpay for the drink, but also that he specifically planned to do so if she didn't allow him to underpay. And the Kantians we are discussing will insist that it is plausible to judge the gangster's use of the barista, that is, his action, to be wrong, not merely to judge that his attitude towards her is problematic.

Those who agree with orthodox Kantians here would, of course, reject the Modified Actual Consent Account[N]. It implies that the gangster does not treat the barista merely as a means, for the gangster might reasonably believe both that the barista gives her informed, voluntary consent to his use of her and that he has made no attempt to get her to submit to his use of her except through getting such consent.[33] However, those who agree with these Kantians would *not* have analogous reasons for rejecting the Possible Consent Account[N]. It does *not* imply that the gangster escapes the charge of using the barista merely as a means, for it is not reasonable for him to believe that she can consent to his use of her, that is, avert his use of her by withholding her consent to it.

My own view is that this Kantian position goes too far. Attempts, even failed or unnecessary ones, to use another and prevent him from averting this usage are often instances of treating the other merely as a means, as the salesperson case illustrates. However, use of another coupled with a *plan* to prevent him from averting this usage does not always count as treating the other merely as a means. The gangster, even

---

[32] Derek Parfit, *On What Matters*, 216.

[33] But the Actual Consent Account[N] might be revised in such a way that it avoids implying that the gangster does not treat the barista merely as a means. According to the relevant elements of the revised account, an agent does not treat another merely as a means if it is reasonable for her to believe that the other gives his informed, voluntary consent to his use of him and the agent *would never* try to get him to submit to her using him through means other than giving his informed, voluntary consent to her using him.

acting on the plan, if necessary, to force the barista to give him a drink does not use her merely as a means if he makes no attempt to force her to give it to him, in my view. That it is reasonable for the gangster to believe that the barista *can share* his end of buying coffee from her is a source of credibility for this view. The gangster's attitude toward the barista is bad, and his character is presumably bad as well, but he does not act wrongly. However, suppose events transpire differently: the barista refuses to sell the gangster coffee for less than its posted price, and he grabs her by the collar to frighten her into giving him the coffee. It would not be reasonable for him to believe that she can share his end of her being frightened. He treats her merely as a means, it seems reasonable to conclude; and that is what the Hybrid Account implies.

4.6 Just Using and Inducing Vulnerability

Although the Possible and Actual Consent Accounts[N] do not seem to be seriously threatened by orthodox Kantianism, both suffer from a significant shortcoming. Since this difficulty has not been appreciated, it makes sense to investigate it in detail, beginning with a case.

A customer has called a technician to his house in order to ready his new computer for use. But the technician intentionally leaves his machine vulnerable to malware in an effort to get business repairing it in the future. A few weeks later, the technician returns to the customer's house. She has learned from him that he aims today to email a document, the only copy of which is on his machine. The document must arrive at its destination today, or the customer will be laid off his job. But his computer is frozen and she is the only one in a position to fix it. The technician uses him to make a profit by getting him to authorize her to do the repair at her usual fee. I think we would say that in getting the customer to authorize the repair, the technician is treating him merely as a means.

However, the Actual Consent Account[N] seems to imply otherwise. We can easily imagine the customer giving his voluntary, informed consent to the technician on her second visit to repair his computer for her usual fee. The account requires that the consent be based on a reasonably accurate understanding of which aspect of the agent will be used (namely his capacity to authorize the repair) and to what proximate end (namely for profit). But the customer's consent might, of course, fulfill these criteria without it being given against the background of information regarding how he came to need to have his computer repaired in the first place. Of course, on the previous service call the customer presumably did not give his informed consent to paying the technician for her work (e.g., because she misled him about what she did and its likely effects). So the Actual Consent Account[N] allows us to conclude that the technician treated him merely as a means *on that occasion*. Yet, implausibly, it seems to imply that she does not do so on the second occasion. This implication is implausible because it seems to be a hallmark of one's "just using" another that one behave as the technician does, that is, that one try to profit from a vulnerability in another for which one bears responsibility.

Defenders of the Actual Consent Account[N] might respond to this criticism in various ways. First, they might claim that, on her second visit, the technician is coercing the customer into paying for a repair. Therefore, he does not voluntarily consent to pay for it. But this claim is implausible. A person counts as coercing another to do something only if he threatens to make the other worse off if he fails to do it, according to the view we earlier attributed to the defenders (e.g., "If you don't give me your wallet, I'm going to shoot you"). But the technician does not threaten to make the customer worse off if he fails to go through with the transaction; she does not threaten him at all.

Second, a defender of the account might claim that the customer does not really give his informed consent to the technician when she visits him for the second time. The customer would be in a position to give informed consent only if he understood the technician's role in bringing about his need for service. In general, the defender might hold, a person can give his informed consent to an agent's using him only if he is informed about her role, if any, in putting him in a situation in which it might further his interests to consent to the agent's using him.

But this expanded notion of informed consent is implausible. Suppose, for example, that a composer happens to see a performance by a raw, but talented, young pianist. Unbeknownst to the pianist, the composer sends a message to his friend, an administrator at a top music school, recommending that the school offer the pianist a scholarship. The school does so, and the pianist emerges from it years later as a promising young professional, an event that would not have occurred had the composer not written that message. The composer, who is now world-renowned, proposes that the pianist play, without pay, in a charity performance of one of his recent works. He wants to use her as a means to making the performance a success. The pianist finds the offer attractive: the event will bring her needed exposure. Now it seems absurd to hold, as a defender of the expanded notion of informed consent must, that the pianist would not be in a position to give her informed consent to the composer's proposal to use her for the performance unless she was made aware of the composer's earlier role in promoting her career. Although the composer himself might never have known, or have forgotten, it is true that he helped put her in the situation in which her allowing herself to be used by him would be to her advantage. But in asking whether she can give her informed consent to perform, we are, intuitively speaking, asking whether she understands how the composer proposes to use her, for example, which work he would like her to play. We are not asking about his role in making it the case that performing might be attractive to her.

We could, of course, multiply examples like this one. What they point to is that, according to reflective common sense, a person can have enough information about a proposed use of him by another to give his informed consent to it, even if he does not have awareness of the agent's role in putting him in the situation in which it might be in his interest for her to use him.

But suppose that we ignore this point and adopt the robust notion of informed consent. It remains implausible to hold that an agent avoids treating another merely as a means if the other gives his informed consent to be used by her. Consider again the example of the computer repair. Suppose that the customer finds out that, on the initial service call, the technician had intentionally left his machine vulnerable to malware. Aware of this, the technician nevertheless demands on her second service call that the customer pay her the usual fee to do a repair. The customer might nevertheless consent to this, that is, to be treated as a means to the technician's making a profit. If he does, he gives his informed consent, even on the expanded notion of such consent. However, it still seems plain that the technician is "just using" the customer and thereby acting (*pro tanto*) wrongly.

A defender of the Actual Consent Account[N] needs to find a path to the conclusion that, when the technician repairs the customer's computer on her second visit, she does not escape the charge of treating him merely as a means. The defender might insist that, despite what it appears, the customer does not on that visit consent to the technician's use of him. Our original description of the case implies that the technician uses the customer to make money twice: once on the initial service call and again when she repairs his machine, giving him access to his document. But the defender might insist that the technician's use of the customer actually constitutes one complex action that has two parts. One part occurs on her initial visit when she attempts to make it the case that the customer will need her in the future; the other part occurs on her second visit when she does the repair and profits from this need. By describing the technician's use of the customer in this way, the defender wants to be in position to say that the customer never gives his informed, voluntary consent to the technician's use of him; for he never gives such consent to being put by her in a position such that he will likely need her services in the future. It is evident that the customer never gives his informed, voluntary consent to this even if he finds out what the technician did to him on her first visit and, on her second visit, nonetheless agrees to her repairing his machine. So, concludes the defender, the Actual Consent Account[N] avoids the implausible implication that the technician is not on her second visit treating the customer merely as a means.

A couple of responses to the defense seem to be in order. First, the defense makes the Actual Consent Account[N] much more complicated to employ in practice than one might have suspected. In order to determine whether someone has given his informed, voluntary consent to being used by an agent, we need to know what this use is. And we cannot determine what it is simply by looking at what the agent is doing to the other at one particular time in pursuit of one particular proximate end. So, for example, suppose that a government establishes a regulated market in human kidneys and sets itself up as the sole buyer. In purchasing a kidney from someone, that is, a vendor, the government is using him—for example, in order to save the life of another citizen. But, according to the defender, we cannot conclude from the fact that the vendor under-stands all the potential risks (e.g., surgical complications) and benefits (e.g., money) of selling his kidney and sells it voluntarily that the government has not treated him merely as a means. In order to reach this conclusion we would need to know whether the government's use of the vendor was a complex action, one part of which was its buying the vendor's kidney, but another, earlier part of which was, say, an effort to convince citizens through a public relations campaign that selling a kidney is a noble thing to do. Suppose that the government's action was complex in the way described and that the vendor's agreement to sell his kidney was conditional on his having been convinced by the government that doing so was noble. We would have license to conclude, based on the Actual Consent Account^N, that the government did not treat the vendor merely as a means when it purchased his kidney only if we had good reason to believe that the vendor gave his informed, voluntary consent to being subject to the public relations campaign. In sum, the proposed defense of the Actual Consent Account^N introduces a way, in addition to those suggested by O'Neill, in which the Actual Consent Account^N is hard to apply.

In any case, I believe that both the actual and *modified* accounts also implausibly imply that the technician does not treat her customer merely as a means, if we alter the example so that it is the following: On her initial service call, the technician is hurried and cuts corners in readying the customer's new machine for use. She foresees that her cutting corners will leave his machine vulnerable to malware and that she might be needed to repair it in the future. Although she says nothing about this to the customer, it is not her aim to set the customer up so that she can profit from him later. After a week, events transpire as they do in the initial version of the example. The technician learns that the customer is desperate to email a document and that she is needed to make this possible. She then uses him to make a profit by getting him to authorize her to do the repair at her usual fee.

In this second case, it seems that she treats the customer merely as a means by profiting from a vulnerability in him that she has foreseeably contributed to bringing about. Yet the accounts imply, implausibly, that she does no such thing; for the customer gives his voluntary, informed consent to serve as a means for her to make a profit. Here a defender of even the Modified Actual Consent Account^N has no grounds to describe the technician's use of the customer in terms of a complex action composed of an attempt to set the customer up to need her services in the future and an attempt to profit from this set-up. The technician simply never attempts to set the customer up.

Based on this sort of example, we might propose that in cases where an agent uses another, she treats the other merely as a means if she is aware that something she has done but could have avoided doing to the other has contributed, in a way foreseeable to her, to making it the case that unless she uses the other, he will suffer a significant reduction in well-being. The technician is aware that she bears responsibility for the agent's needing to use her in his effort to keep his job. Moreover, focusing on the latter variation on the example, the technician realizes that it was her choice to cut corners and leave the customer's computer vulnerable to malware and that he might have an urgent need for her to get the malware off his machine in the future.

But this proposal suffers from a serious flaw, which is easy to illustrate with an example. In order to make money, a plastic surgeon has pioneered an operation that might make a patient look younger. The patient understands the significant risks that the operation carries, one of which is that the surgeon might have to perform a second surgery. The surgeon will not operate on the patient at all unless the patient consents to these risks, and the patient does so. A few days after the procedure, something goes wrong and the patient's life is in jeopardy. The only one who is in position to save him is the surgeon, by operating again. The account implies, implausibly, that in operating again the surgeon is treating the patient merely as a means. For something she has done to him, namely perform the first operation, has contributed to making it the case that, unless the surgeon uses the patient, that is, does the second operation, his well-being will diminish significantly. When a person has had the opportunity to assume the risk that an agent's use of him will create a need in him for this very person to use him again, it can be implausible to conclude that the agent is "just using" the other.

This shortcoming as well as others in the proposal might be rectified with a more sophisticated one: the Induced Vulnerability Account.

Suppose an agent uses another. She uses him merely as a means if:

1. It is reasonable for the agent to believe that something she has done to the other has contributed to his being in the position that unless the agent herself uses him, his well-being will diminish significantly.
2. It was foreseeable to the agent that she would contribute to his being in this position as well as avoidable that she would do so.
3. The agent was, as a practical matter, able to but did not give the other an opportunity to dissent from taking the risk that she (the agent) would contribute to the other's being in this position.
4. The end of the agent's use of the other is not limited to her discharging what she reasonably believes to be a moral obligation towards him.

Some elements in this account require clarification. In accordance with previous usage, the account invokes a non-moral notion of reasonableness. It is reasonable for an agent to believe something if his believing it is justified in light of his epistemic situation. It is presumably reasonable for the plastic surgeon to believe that something she has done to the patient, namely operate on him, has contributed to his being in the position that unless she operates again his well-being will diminish significantly. For if she does not operate again, he will die. Of course, people have various views on what constitutes well-being. They disagree about whether it consists in experiencing pleasure, satisfying one's desires, realizing a set of goods, perhaps including friendship and knowledge, or some combination of these. The Induced Vulnerability Account invokes what it is reasonable for *the agent* to believe regarding the other's well-being. It evaluates the agent's action in terms of her own views regarding the other's well-being, including what his well-being consists in, as long as these views are reasonable.

The second part of the account appeals to notions of foreseeability and avoidability. The effects of an agent's action are foreseeable or avoidable for her only if, as a practical matter, she is able to foresee them or avoid them. Although the patient's postoperative complication was unlikely, the surgeon was able to foresee it, and she could have avoided it by refraining from carrying out the operation. However, it would not have been foreseeable to the surgeon that, say, the patient would meet his future spouse at her office.

The notion of an opportunity to dissent, invoked in the third part of the Induced Vulnerability Account, is familiar from our discussion of possible consent. To have an opportunity to dissent to being used is not merely to be able to express one's dissatisfaction with it, but to be able to avert the use by refraining from agreeing to it. The patient in our example had such an opportunity; for he was both informed of the risk that the initial operation would make it the case that unless he had a second one, his health would deteriorate, and he was free to refrain from agreeing to undergo the initial operation, thereby precluding the surgeon's use of him. So this account does not imply that the surgeon is "just using" the patient. Of course, since the surgeon gave the patient the opportunity to dissent, she was able, as a practical matter, to do so. An agent is able, as a practical matter, to give another such an opportunity if the agent and the other's circumstances are compatible with her doing so. That it might jeopardize an agent's profits to give another an opportunity to dissent from taking the risk that he might need her services to avoid future loss of well-being is, of course, compatible with its being practically possible for her to give her such an opportunity.

Finally, in order to fulfill the sufficient condition for using another merely as a means contained in the Induced Vulnerability Account, it must not be the case that the agent is using the other in order to discharge what she reasonably takes to be a moral obligation to him. Consider the second version of our example, namely the version in which the technician foreseeably but not intentionally leaves the customer's computer vulnerable to malware. When the desperate customer calls her back for a repair, she might not charge him at all, in order to make up for the shoddy work she had done earlier. She would be using the customer to even out the moral balance sheet, but it seems implausible to say that in doing that, she would be treating him merely as a means. The final condition in the account allows it to avoid this implausible implication.

The account strives to crystallize *one set* of jointly sufficient conditions under which an agent treats another merely as a means. It generates normative verdicts that clash with those yielded by the Modified Actual Consent Account[N]. Whereas the Induced Vulnerability Account implies, plausibly, that, in the second version of our example, the technician treats the customer merely as a means, the Modified Actual Consent Account[N] implies, implausibly, that she does not. I take this to be a noteworthy shortcoming of the latter account.

But it is important to observe that the Possible Consent Account[N] fares no better. According to the account, let us recall, an agent does not treat another merely as a means if it is reasonable for the agent to believe that the other can consent to her use of him. The other can consent to her use of him if he can avert this use by withholding her consent. On the technician's second visit, it is reasonable for her to believe that the customer can avert her use of him in this way. For, as she knows, all he needs to do is to tell her "No thank you" and she will leave.

We might be tempted to maintain that it is actually *not* reasonable for the technician to believe that the customer is able, by withholding his consent, to avert her use of him to make a profit. After all, as the technician is aware, the customer needs to get his computer repaired to keep his job. However, this reasoning conflates the impossibility of someone's averting an agent's use of him with the undesirability from his perspective of his averting this use. The customer has the power to prevent the technician's use of him, albeit at the risk of his job, in a way that victims of coercion and deceit do not. As the technician knows, the customer can stop her from using him simply by not consenting to hire her. The victim of a mugging can withhold his consent to being used by the mugger all he wants. But the mugger is going to use him in order to get money, regardless of whether he does it the "easy way or the hard way." So, in sum, it is reasonable for the technician to believe on her second visit that the customer can consent to her use of him. Of course, we might respond by modifying the Possible Consent Account[N] so that a person can avert an agent's use of him only if his disallowing this use would not result in a decrease in his well-being. But this modification would have a high price. For example, it would imply that someone who needs heart surgery in order to survive and enjoy a good life cannot consent to the operation.

The plausibility of a possible consent or an actual consent account of an agent's using another, but not merely as a means, increases if it avoids generating results that clash with the Induced Vulnerability Account. The Final Possible Consent Account[N] tries to accomplish this:

Suppose an agent uses another. She does not use him merely as a means if it is reasonable for the agent to believe that:

the other can consent to her use of him
and
nothing she has done to the other has contributed to his being in the position that unless the agent herself uses him, he will undergo a significant loss of well-being.

Of course, one might amend the Modified Actual Consent Account[N] in an analogous way. According to the Final Actual Consent Account[N]:

An agent uses another, but does not use him merely as a means, if it is reasonable for her to believe that
the other gives his voluntary, informed consent to her using him,
she has made no attempt to get him to agree to her using him other than that of trying to gain his voluntary, informed consent to her using him, and

nothing she has done to the other has contributed to his being in the position that unless the agent herself uses him, he will undergo a significant loss of well-being.

It would be remiss not to acknowledge that both the Final Possible and the Final Actual Consent Accounts[N] are subject to an objection. Echoing a case discussed earlier, suppose that a patient with a serious heart condition needs and wants an operation immediately in order to survive. A surgeon, who has never before interacted with the patient and who is the only one who can perform the surgery, proposes to operate not at her typical, already considerable fee, but for an even higher one. She informs him of the risks of the operation and so forth, and makes it clear to him that unless he pays her fee, which she acknowledges to be higher than usual, she will not operate. The patient agrees, and the surgeon uses him to make a profit, saving his life in the process. The Final Actual Consent Account[N] implies that the surgeon does not treat the patient merely as a means. For it is reasonable for her to believe that the patient gives his voluntary, informed consent to her using him; she has not tried to get him to agree to the operation through means other than asking for this consent (e.g., she has not employed deception), and nothing she has done to him has contributed to his being in a situation in which her use of him is necessary to prevent a significant loss of well-being. The Final Possible Consent Account[N] also implies that the surgeon does not treat the patient merely as a means. For the patient can avert the surgeon's use of him by withholding his consent to it.

How implausible are these implications? Of course, the accounts do not aim to capture the whole of morality. They try to specify conditions under which someone uses another, but does not treat him merely as a means. But someone can act wrongly with respect to another without using him at all, for example by behaving toward him as one would behave toward an inanimate obstacle to be kicked away. Moreover, to say that, if certain conditions are met, someone is not using another merely as a means fails to imply that the person's *use* of the other is morally permissible. The accounts leave open the possibility that the surgeon's use of the patient is wrong because it fails to accord with her professional obligations as a licensed physician, or because it discriminates unfairly between this patient and others whom she charges her normal fee, or, perhaps most obviously, because it exploits the patient in the sense that it takes unfair advantage of him.

Both the Final Actual and Final Possible Consent Accounts[N] imply that some cases of what we might intuitively label as an agent's taking unfair advantage of another are not also cases of treating the other merely as a means. They might, therefore, fail to capture the full range of cases in which some find it natural to say that someone is "just using" another.

But I doubt whether any philosophically satisfying account of not treating others merely as means is going to square entirely with ordinary judgment. We might try to build into an account a condition such that if an agent takes unfair advantage of another, we must refrain from concluding that she is not treating the other merely as a means. But such an account would not be very illuminating. The notion of taking unfair advantage of another is closely related to that of just using another. (We might describe the technician's behavior towards her customer either as her just using him or as her taking unfair advantage of him, for example.) Moreover, the notion of taking unfair advantage of another is no less in need of clarification than that of treating another merely as a means. For example, suppose that the patient is impoverished. Does the surgeon take unfair advantage of him if she charges her usual fee, or half of her usual fee, or less than that, but more than he can easily afford? It would be easy to multiply such questions. In any case, it is hardly surprising that rendering precise a commonsense idea of treating others merely as means involves some narrowing of the idea's scope. The accounts our discussion has generated might not accord fully with ordinary thinking. But they nevertheless promote understanding of the notion of treating others merely as means.

The Actual Consent Account[N] with which we began this chapter has strong initial appeal. If someone gives his informed, voluntary consent to serve as a means to some end, then how could it be that he is being treated merely as a means? After all, he agrees to be used. This chapter has tried to show the account's appeal to be questionable. Someone's agreeing to be used by an agent is compatible with her using him merely as a means. This compatibility can be manifest when the agent's use of another involves her profiting from some vulnerability in him that she has helped to induce. It can also be evident when the agent's use of the other involves an otiose attempt at deceiving or coercing him. We can amend the Actual Consent Account[N] so that it is sensitive to the possibility of an agent's treating another merely as a means in these ways. But the necessary changes yield an account that is both considerably more complex than the Actual Consent Account[N] and that, for practical purposes, differs little from possible consent accounts.

5

Dignity and the Mere Means Principle

This chapter culminates in a new account of the dignity of persons. Part of what constitutes their dignity is their having a status such that they ought not to be treated merely as means. So the chapter begins by filling out our understanding of treating others merely as means. Thus far, we have championed two sufficient conditions for an agent's using another merely as a means: the Hybrid and Induced Vulnerability Accounts. We have also embraced as plausible two sufficient conditions for an agent's using another, but *not* merely as a means, namely the Final Possible Consent and the Final Actual Consent Accounts[N]. In Chapter 4 we discussed at length accounts of using another, but not merely as a means that incorporate a notion of actual consent. But we have yet to consider actual consent accounts of sufficient conditions for treating another merely as a means. This chapter tries to show (5.1) that such an account would, like the Hybrid Account championed in Chapter 3, have to be agent-focused and complex. In order to make our account of the Mere Means Principle more complete, the chapter also specifies a *necessary* condition for an agent's treating another merely as a means (5.2). With the aim of solidifying our understanding of some of our accounts of treating others merely as means, the chapter then applies them (5.3) to stylized cases involving transplant surgeons, runaway trolleys, and so forth. These applications should reinforce the accounts' plausibility.

The chapter next develops an account of dignity: one that incorporates our specification of the Mere Means Principle (5.4). The account holds that dignity is a special status held by persons. In order to clarify the account as well as to underscore its plausibility, the chapter revisits the examples that, according to Chapter 2, diminished the credibility of a more traditional Kantian account of dignity (5.5). The new account sets forth a necessary condition for an agent's respecting a person's dignity. It is not a complete account of the dignity of persons. Although it points to some actions that fail to respect their dignity, it does not try to specify all such actions.

At the end of the chapter (5.6), we revisit an issue addressed briefly in Chapter 3. There we endorsed the view that the Mere Means Principle is most plausibly understood to express an overridable constraint, that is, to hold that it is always *pro tanto* wrong to treat another merely as a means, but not necessarily wrong all things considered. But that view prompts the question of when treating another merely as a means is morally permissible. With the introduction of our Wrong-Making Insufficiency Principle, we offered a partial answer to that question. An analogous question arises in connection with the account of dignity that the chapter sketches. The chapter maintains that it is always *pro tanto* wrong to fail to respect the dignity of persons, but it is not always wrong all things considered. The chapter offers some conjectures as to when it is morally legitimate to fail to respect some person's dignity.

5.1 Actual Consent and Treating Another Merely as a Means

Our initial task is to complete unfinished business regarding the Mere Means Principle. We need first to discuss the prospects of developing a sufficient condition for treating another merely as a means that incorporates a notion of actual consent. A proposal for such a condition is not hard to formulate. We might say that an agent who uses another uses him merely as a means if the other has given his voluntary, informed *dissent* to her use of him. For someone to offer dissent to an agent's use of him is just for him to express that if he had the power to determine whether this use of him would take place, given all that has transpired to this point, it would not. It does not suffice to conclude that a person in need of a life-saving operation dissents from it that he would prefer not to need any operation at all. We can understand dissent being *voluntary* and *informed* in the same senses we specified above (4.1). Dissent is voluntary only if it does not result from someone's threat to make the dissenter worse off than he would be if he never interacted with that person. Someone's dissent to being used by an agent is informed if he understands what the agent intends to do to him, with which likely effects, and to what proximate end. This proposal has two features in common with its cousin, the Actual Consent Account[N] of using another, but *not* merely as a means. First, it seems to have the virtue of simplicity. Second, it is patient-focused. In determining, based on it, whether an agent is treating another merely as a means, we concentrate on what the other (the patient) has done vis-à-vis the agent's using him. But like its cousin, this proposal is implausible and must be modified so that it comes to resemble the Hybrid Account in being agent-focused and complex.

The proposal faces an immediate difficulty that stems from the simple fact that people change their minds. For example, someone has given his voluntary, informed consent to a photographer to serve for a fee as her model for an hour at her studio. But after half an hour has elapsed, he pockets his fee and moves towards the exit, proclaiming that he has better things to do. He has now given his voluntary, informed *dissent* to the photographer's continued use of him. However, it would be implausible to agree with the proposed account that if she takes a couple of photographs of him on his way out, she treats him merely as a means.[1] In some cases, when an agent uses

---

[1] Or suppose that during the American Civil War a suffering Union soldier hires a surgeon he comes across in the South to amputate his leg. He might tell the surgeon in advance to go ahead with the operation,

another in a way they both previously agreed to, the agent is not, intuitively speaking, treating the other merely as a means, even if the other now dissents from this use.

A further shortcoming of the proposal is that it fails to capture a swathe of cases where, intuitively speaking, someone is treating another merely as a means. It fails, for example, to imply that the desperate salesperson is treating the office manager merely as a means when she threatens to reveal his sexual preference if he does not give her the sales leads (4.4). The office manager does not dissent to her using him to get the leads. In general terms, the proposed account fails to designate as an agent's treating another merely as a means her trying to make her use of the other successful with the help of otiose or ineffectual threats to make him worse off unless he does what she wishes. It also fails to entail that an agent is just using another when her usage of the other is predicated on her deceiving him. A crooked auto mechanic who for profit deceives a customer into authorizing the installation of a new transmission might reasonably claim that the customer never gave his voluntary, informed *dissent* to her use of him; for she was unaware of what that use was so did not have an opportunity to express her dissent.

These difficulties might be overcome. But the language necessary to do so without circularity is cumbersome:

Suppose an agent uses another. She uses him merely as a means if it is reasonable for her to believe that
either

    a) the other was informed, before it occurred, of the agent's intended use of him and at that time voluntarily dissented from it, or

    b) the other was not or could not, before it occurred, be informed of her intended use of him. But if the other had been so informed, he would have voluntarily dissented from it,

and

    c) the other has not, prior to the agent's use of him, given his voluntary, informed consent to it or to a set of rules governing his and the agent's interaction, according to which her use of him is legitimate.

Let us call this the Actual Consent Account, in contrast to the Actual Consent Account^N, which presents a sufficient condition for an agent's using another, but *not* merely as a means. We will explain the account with the help of the cases we just considered.

We notice immediately that this account is more agent-focused than the initial proposal. Its conditions are stated in terms of what it is reasonable for the agent to believe regarding his use of the other. This agent focus enables the account to register that, as many of us hold, the desperate salesperson is treating the office manager merely as a means. According to the case as we have envisaged it, it is reasonable for the

---

even if he pleads with him to stop, for the soldier knows that amputation is his only hope of survival. If the surgeon continues with the operation, despite the soldier's at one point pleading for him to stop, he does not treat him merely as a means.

salesperson to believe that if the manager had been informed of her intended use of him to get sales leads before it occurred, he would have dissented from it. And the salesperson is unaware that, according to company policy, the manager is to provide her with leads upon request. The revised account also implies plausibly that the crooked mechanic was treating her customer merely as a means. If the customer had been informed of the mechanic's intended use of him, namely to get him to authorize the unnecessary installation of a new transmission, he would have dissented from it, it is reasonable for the mechanic to believe. Moreover, it was not reasonable for her to believe that the customer ever gave his informed consent to this unnecessary installation or to a set of rules according to which it was legitimate.

Condition c) prevents the account from having implausible implications in cases like that of the photo shoot. The model did give his voluntarily, informed agreement to being used as the photographer uses him. This condition also prevents the account from having another implausible implication. Suppose that in order to better his ranking, a professional tennis player goes ahead and defeats a desperate competitor despite the competitor's dissenting from his using him in this way. (We can imagine the competitor whispering to him at changeovers: "Please, just let me win this one match.") The clause enables us to avoid the conclusion that the player is treating his competitor merely as a means; for it is reasonable for the player to believe that the competitor has given his voluntary, informed consent to a set of rules, namely those of the Association of Tennis Professionals, according to which the player's use of him is legitimate. Finally, condition c) insulates the account against implying that a patrolman is using a speeding motorist merely as a means in giving him a ticket to make her monthly quota. When they apply for their drivers' licenses, motorists presumably give their voluntary consent to a set of rules according to which an officer can legitimately ticket a motorist who is speeding, regardless of whether she is doing so to meet her monthly quota.[2]

With the help of a notion of actual consent, one might hope to generate a simple and plausible patient-focused sufficient condition for an agent's treating another merely as a means. But a plausible actual consent condition turns out to be neither patient-focused nor simple. Whether we are inquiring into sufficient conditions for using another but not merely as a means as we did in Chapter 4 or sufficient conditions for using another merely as a means as we have been doing here, we find that plausible actual consent accounts, like possible consent accounts, end up being complex and agent-focused.

[2] Like the Hybrid Account, the Actual Consent Account implies that the officer in the case of Arrest (3.5), namely the one who is using the white supremacist as a source of information, is treating him merely as a means. In defending the Actual Consent Account we would presumably embrace the idea that treating another merely as a means is wrong *pro tanto*, but not necessarily wrong all things considered. We would then develop a principle such as our Wrong-Making Insufficiency Principle (3.5) in order to highlight why, in cases such as Arrest, treating another merely as a means is not wrong, all things considered. But I will not try to develop such a principle here.

In Chapter 4 we also found that plausible actual and possible consent accounts of using another but not merely as a means overlap: cases in which an agent's use of another fails to fulfill the conditions specified in one account also tend to be cases in which it fails to fulfill those specified in the other. A similar point applies regarding accounts of treating others merely as means, namely the Actual Consent Account, which we just developed, and the Hybrid Account. Focusing on condition b) in the Actual Consent Account, for example, suppose it is reasonable for an agent to believe that had another been informed of the agent's use of him prior to its taking place, the other would have voluntarily dissented from it. It is reasonable for Sue, a surgeon, to believe that Bob, a patient visiting the hospital for routine tests, would not have agreed to her using him as she did, namely as a source for the heart and lungs necessary to save two strangers in need of transplants. It would typically also be reasonable for the agent to believe that the other could not share the end she was pursuing in using him in the sense specified in the Hybrid Account. For it would be reasonable for her to believe that his dissent would have stemmed from his holding both that being used by her would interfere with his attaining his own ends and that the likely effects of her using him would not prompt him to give up these ends. It would be reasonable for Sue to believe that Bob's dissent would stem from his holding both that being used by her would preclude him from realizing his aim of staying alive and that he would not be willing to relinquish this aim, even in the light of the likelihood that the removal of his heart and lungs would lead to the preservation of two lives. If we add that Bob had no opportunity to avert Sue's use of him and that at no point did he agree to the organ extraction or to rules according to which it was legitimate, then we can conclude that Sue was treating him merely as a means, according to both the Actual Consent and the Hybrid Accounts. If there is a gap between the set of cases that the two accounts imply to be ones of an agent's treating another merely as a means, it is a narrow one indeed.

5.2 A Necessary Condition for Treating Another Merely as a Means

In order to further our project of specifying the Mere Means Principle, we need to develop a necessary condition for an agent's treating another merely as a means. Since, in filling out other aspects of the principle, we have not found appealing to actual, as opposed to possible, consent to augment its acceptability or perspicuity, we will limit ourselves to thinking in terms of possible consent. Fortunately, such an account is implicit in work we have already done. Keeping in mind our aim of specifying the Mere Means Principle in such a way that it is plausible to claim that in treating another merely as a means, an agent is *acting pro tanto* wrongly, let us propose the following account:

Suppose an agent uses another. She uses him merely as a means only if it is reasonable for her to believe either:

    a) that the other is unable to consent to her using him

or

    b) that she has contributed to making it the case that the other's well-being will significantly diminish unless she uses him.

Let us call this the Disjunctive Necessary Condition Account. The first condition incorporates the notion of possible consent contained in the Hybrid Account. A person is unable to consent to someone's use of him if he cannot avert this use by withholding his agreement to it. But a) alone does not constitute the necessary condition we seek. An agent can treat another merely as a means, we found, even if the other is able to consent to her use of him. She does so when she fulfills the conditions set forth in the Induced Vulnerability Account, that is, (very roughly) when, in a way foreseeable to her, she helps make it the case that in order to avoid a significant loss of well-being, a person needs to consent to her using him. That a person needs to consent to an agent's using him in order to avoid such a loss does not entail that he is unable to avert this usage by withholding his agreement to it. Suppose that a local guide intentionally leads a hiker into a dangerous situation with the hope of profiting from his hardship. It turns out that unless he pays her to trudge up the mountain and rescue him, he will lose fingers and toes. The hiker would presumably nevertheless have the power to avert the guide's use of him to make money simply by refusing her help. It might be reasonable for an agent to believe that her use of another fulfills condition b), but not condition a). Of course, it might also be reasonable for an agent to believe that her use of another fulfills a), but not b). We can easily imagine a customer who cannot consent to a deceitful auto mechanic's use of him, but who is not in a position such that his maintaining his level of well-being depends on the mechanic's using him. Neither condition a) nor condition b) is plausible as a stand-alone necessary condition for an agent's just using another.

5.3 Applying the Mere Means Principle

Our efforts to specify the Mere Means Principle charitably, that is, so that it is plausible to claim that to treat someone merely as a means is to act *pro tanto* wrongly, has yielded accounts that fall into three categories. First, we have put forward as plausible three sufficient conditions for an agent who is using another to count as using him merely as a means: the Hybrid Account (3.5), the Induced Vulnerability Account (4.6), and the Actual Consent Account (5.1). Second, we have championed two sufficient conditions for an agent who is using another to count as *not* using him merely as a means, namely the Final Actual Consent and Final Possible Consent Accounts[N] (4.6). Finally, we have just embraced a necessary condition for an agent who is using another to qualify as treating him merely as a means: the Disjunctive Necessary Condition Account. Two of the sufficient conditions for treating others merely as means, namely the Hybrid and the Actual Consent Accounts, imply that roughly the same range of actions involve treating others merely as means. The third sufficient condition for treating others in this way, namely the Induced Vulnerability Account, picks out a different range of actions and thus in effect compensates for a shortcoming in scope of the other two accounts. Our two sufficient conditions for using another, but not merely as a means, the one based on possible and the other on actual consent, also coincide in their scope (4.4).

By way of summary, it might be helpful to apply some of the accounts to stylized cases familiar from the literature on moral constraints. For the sake of brevity, let us limit ourselves to one representative of each type of account I have just mentioned: the Hybrid Account, the Final Possible Consent Account[N], and the Disjunctive Necessary Condition Account. Applying the accounts to these cases should help to confirm their plausibility. Some of the cases are familiar from earlier discussions, although they are stated in slightly different terms here:

Rescue: One person is stranded on one island and five people are stranded on another island. Everyone will die soon unless you perform a rescue in your boat. But the islands are so far apart that if you rescue the one, the five will die, and if you rescue the five, the one will die.

Trolley: A driverless runaway trolley is heading towards five people stuck on the track. If not diverted, it will kill all five. By operating a railroad switch, you could direct the trolley onto a different track where it would kill only one person.

Loop Trolley: A driverless runaway trolley is heading towards five people stuck on a track. By operating a railroad switch, you could redirect the train onto another track. But that second track loops around and rejoins the original where the five are stuck. You notice that there is a person on the second track. If you redirect the trolley onto that track, it will run into the person, killing him, but the impact will trigger the trolley's emergency brake so that it does not kill the five. If you do not redirect the trolley, it will kill the five but the impact will trigger the brake so it will pose no threat to the one.[3]

Regarding all of the cases we discuss in this section, let us assume that you, a passerby, neither know nor are able to discover any potentially morally relevant differences between the people, all strangers to you, who are facing possible death; and you cannot communicate with any of the people at all. For example, in Trolley, you have no reason to believe that the one person is a pedestrian crossing the tracks at an officially designated spot, while the five are trespassing vandals. You do not set out in Rescue with grounds for thinking that the five on the one island are all octogenarian prison escapees whereas the one on the other is a 25-year-old surgeon working for a relief agency. Let us also assume that the cases have no morally relevant effects except for those they describe. For example, publicity generated from your refraining from operating the switch in Trolley would not promote the passage of a new "Good Samaritan" law.

The Rescue and Trolley cases are included to emphasize an important point: the accounts have no implications regarding them. Each of the accounts applies only to

---

[3] These cases (or, rather, slight variants of them) are sketched by Judith Thomson. See Thomson, "The Trolley Problem," in *Rights, Restitution, and Risk* (Cambridge, Mass.: Harvard University Press, 1986).

cases in which an agent is *using* another. Regarding Rescue, suppose you decide to save the five. In rescuing them, you would not be using the one in the sense of using we have embraced. You would not take the one's presence or participation to contribute to your accomplishing your goal of getting the five off the island. Of course, if you rescued the one instead, you would not be using the five to get him off the island either.[4] An analogous point applies to Trolley. By operating the railroad switch in order to divert the train away from the five and towards the one, you would not be using the one. For you would not intend his presence or participation to help you to accomplish your goal of saving the five. As far as your accomplishing it is concerned, it would make no difference whether he was there or not. Moreover, in refraining from throwing the switch, you would obviously not be using the five to save the one. If your saving some set of persons rather than another set in Rescue and Trolley is morally impermissible, its impermissibility does not stem from its amounting to your using anyone who is not saved merely as a means. For no matter whom you save in these cases, in so acting you would not be *using* the person or persons you do not save. Of course, the view that saving some rather than saving some other(s) in these cases does not involve treating anyone merely as a means does not entail that saving some rather than some other(s) is morally permissible.

In the Loop Trolley case, if you do not engage the switch, but instead, say, call the authorities, you are thereby neither using the one nor the five. I take it to be obvious that you are not using the one. You are not intentionally doing anything to him. Moreover, that you allow the five to be killed does not entail that you are using them. For you presumably do not intend their presence or participation to contribute to any end you are pursuing. Our accounts thus do not entail a verdict regarding the moral status of your not engaging the switch.

However, if you save the five, then you *do* use the one. For you intentionally do something to him, namely divert the trolley towards him, in order to save the five, and you intend him to contribute to your end's realization by slowing the trolley down. What do the accounts imply regarding your use of him? This use satisfies the Disjunctive Necessary Condition for treating another merely as a means. As you are aware, you cannot communicate with the one in any way. So it is obviously reasonable for you to believe that he cannot avert your use of him by withholding his agreement to it. Moreover, since it is reasonable for you to believe this, your action of engaging the switch to save the five would not satisfy the Final Possible Consent Account[N] for using another but not merely as a means. In order to satisfy that account, it must be the case that it *is* reasonable for the agent to believe that the other can consent to her use of him.

Would your action fulfill the sufficient condition for treating another merely as a means detailed in the Hybrid Account? Since we have already found that the one is unable to consent to your using him, the remaining question is whether it is reasonable

---

[4] If you rescued no one, you would, of course, also be using no one.

for you to believe that the one cannot share the (proximate) end you are pursuing in using him. Recall that a person cannot share an agent's end if and only if: The person has an end such that his pursuing it at the same time that he pursues the agent's end would violate the hypothetical imperative, and he would be unwilling to give up pursuing this end, even if he was aware of the likely effects of the agent's pursuit of her end and aware that, based solely on his preference, the agent would give up her pursuit of her end. It would be reasonable for you to believe that the one has an end, namely any end the realization of which requires that he be alive, such that his pursuing your end of his body's slowing down the trolley would preclude him from realizing his end and thereby amount to his violating the hypothetical imperative.

But is it reasonable for you to believe that the one would be unwilling to give up pursuing an end of his such that his realizing it requires that he be alive, even if he was aware that his body's slowing down the trolley would save five people and that your knowledge of a preference of his not to have his body used to slow it down would suffice for you to refrain from using it in this way? Strictly speaking, the example is too artificial to warrant a confident answer. The answer would depend on what it is reasonable for you to believe regarding the willingness of the one to die so that others, the identity of whom he is ignorant, may live. That in turn would depend on your beliefs regarding the one's motivations, which might vary according to the cultural context. We can at least imagine societies in which willingness to sacrifice one's own life to save others is the norm. But in many contexts, it seems obvious that it would be reasonable for you to believe that the one would be unwilling to sacrifice his ends, including the simple end of going on living, for the sake of saving the five. We tend to call those who are willing to make such sacrifices heroes, a description that presumably fails to apply to most of us. Whereas in Trolley, if you save the five you are not using the one at all, in Loop Trolley you are, in many contexts, treating him merely as a means, according to the Hybrid Account.

When she introduces Loop Trolley, Judith Thomson suggests that it would be implausible to hold it to be morally impermissible to operate the switch in that case, but permissible to operate it in Trolley. "[W]e cannot really suppose that the presence or absence of that extra bit of track makes a major moral difference as to what an agent may do in these cases," she says.[5] Extra track does form the loop. But it is not the extra track that marks the moral distinction between the cases, in my view. Rather it is the fact that in Loop Trolley, you treat the one person merely as a means, but in Trolley you do no such thing.[6]

---

[5] Thomson, "The Trolley Problem," 102.
[6] According to my view, that you treat the one merely as a means in Loop Trolley entails that you act *pro tanto* wrongly. I also believe that you act wrongly, all things considered. Some philosophers share this belief (e.g., Henry Richardson, "Discerning Subordination and Inviolability: A Comment on Kamm's *Intricate Ethics*," *Utilitas* 20 (2008): 88), while others do not (e.g., Frances Kamm, *Intricate Ethics: Rights, Responsibilities, and Permissible Harm* (Oxford: Oxford University Press, 2007), 92).

For the sake of brevity, I will not do so, but it would be easy to show that the results of applying our three accounts to the Loop Trolley case would be mirrored in applying them to two further familiar cases:

Bridge: A driverless runaway trolley is heading towards five people stuck on a track. You could save the five only by using remote control to make a person standing on a bridge above fall in front of the trolley, thereby killing him, but also triggering the train's automatic brake so that it does not kill the five.

Transplant: A healthy person is in the hospital for a study on a new sleeping pill. You are a surgeon who knows that if you approach him in the sleep ward and extract his organs, thereby killing him, you can use them to save five others who would otherwise soon die.

If you refrain from saving the five in either case, you do not use the five or the one, so the accounts fail to apply. Saving the five would satisfy the Disjunctive Necessary Condition for the agent's treating another merely as a means, as well as the sufficient condition for doing so specified in the Hybrid Account. But it would not fulfill the condition for using another, but not merely as a means, detailed in the Final Possible Consent Account^N.

5.4 A Kant-Inspired Account of Dignity (KID)

A second main aim of this chapter is to sketch an account of the dignity of persons and what it means to respect it that has more plausible normative implications than the Respect-Expression Approach to the Formula of Humanity (FH) that we probed in Chapter 2. According to that approach, FH amounts to RFH: *Act always in a way that expresses respect for the worth of humanity, in one's own person as well as that of another.* RFH is to be understood as the supreme principle of morality, that is, as a principle that all of us have an overriding obligation to conform to and as the norm from which all genuine moral duties derive. According to the Respect-Expression Approach, to say that persons have dignity is to say that they have unconditional and incomparable worth: a worth not to be maximized, but rather to be respected. The unconditional worth of a person is a worth beyond price: one that is neither greater nor less than that of any other person and that does not diminish no matter what she does or what happens to her. The incomparable worth inherent in a person is a worth that it is never legitimate to exchange, even for the worth inherent in several other persons. An agent's action respects the dignity of persons if and only if it expresses proper respect for this worth. In order to do that, the action must not send a message that is incompatible with the notion that persons have unconditional and incomparable worth.

We have found that this approach has implausible implications in various cases. To recall one, contrary to what many of us believe, the approach implies that a soldier acts wrongly if he intentionally kills himself by jumping on a grenade in order to save the lives of four others. That action sends the message that the one person's worth is not as great as that of the four taken together, thereby expressing disrespect for the *incomparable* worth of his humanity. In order to remedy this problem, one might be tempted to change the approach so that it holds that persons have unconditional but not incomparable worth. Worth that is unconditional, but not incomparable, can be aggregated. If two persons have unconditional but not incomparable worth, then two persons together have more worth than one, it is legitimate to claim. According to this reconstruction, FH would not imply the soldier's action to be wrong. His sacrificing his own humanity in order to preserve that of four others would not send the message that humanity lacks unconditional worth. It would send one consistent with the idea that his own humanity has unconditional worth, just not as much as that of his comrades taken together.

Unfortunately, this change would not be an improvement, many of us believe. Given that the modified approach implies that the one's diving on the grenade was morally permissible, it would imply the same regarding the action of a soldier in similar circumstances who, instead of diving on the grenade himself, shoved an unwilling comrade onto it, thereby saving himself and three others. For in this case like the other the action would send a message compatible with the notion that persons have unconditional worth. It would cohere with the idea that the person he pushed onto the grenade has unconditional worth, albeit less worth than he and the three others together. Of course, the modified approach would also imply that it was morally permissible to use the pedestrian as a trolley decelerator in Bridge, and to use the healthy research volunteer for his organs in Transplant—implications that many of us find counterintuitive.

We can avoid these implications if we hold that the dignity of persons lies not only in their having unconditional value, but also in their being such that it is wrong to treat them merely as means, according to the accounts of doing so that we have specified. According to the Hybrid Account, for example, the soldier who pushed his reluctant comrade onto the grenade would be using him merely as a means; it would be reasonable for the soldier to believe neither that his comrade could consent to his use of him nor that he could share the (proximate) end he was pursuing in using him, namely that he (the comrade) absorb the impact of the exploding grenade.

But this approach is also problematic. Modifying an example we discussed at length (2.4), suppose that an innocent journalist trapped in an alley is being attacked not by one but by two former paramilitary soldiers coming at him knife-raised. After brandishing a pistol at them and making an ineffectual demand that they stop, the journalist believes reasonably that in all likelihood the only way to save himself is to kill both attackers. So he shoots both, trying to kill them. Many of us find his action to be morally permissible. But it would not seem to express respect for the dignity of persons, according to our current approach. Granted, the Hybrid Account would not entail that, in intentionally killing his attackers, the journalist was using them merely as means. He would not be using them at all; for he would not intend their presence or participation to contribute to his end of not being stabbed to death. However, recall that, on the current approach, the attackers possess unconditional value. Their murderous endeavors do not diminish this worth one whit. In doing what he reasonably deems necessary to save himself, namely killing the two attackers, the journalist would be implying that he is more valuable than two others. But that message would contradict the notion that each person involved has unconditional worth (as the approach understands it), namely a worth beyond price that is neither greater nor less than that of any other person.

In sum, RFH commands us to act always in a way that expresses respect for the worth of humanity: its dignity. If we specify that worth to be both incomparable and unconditional, we must embrace unwelcome conclusions, including that the soldier who intentionally forfeits his own life to save that of several others acts wrongly. Specifying the worth of humanity to be unconditional but not incomparable allows us to escape this conclusion, but at a cost many of us are unwilling to pay, namely that of having to agree, for example, that it is morally permissible to use the one in Bridge as a trolley decelerator to save the five. Building into the notion of the dignity of persons not only that they have unconditional worth, but also that they ought never to be treated merely as means enables us to avoid this cost, but seems ultimately unsuccessful. The aggregative feature of unconditional worth, as opposed to unconditional and incomparable worth, is a double-edged sword. In the context of the Respect-Expression Approach to FH, it does not allow us to adopt a view many of us find plausible, namely that, in some situations, an agent acts rightly in trying to save himself, even if the means he takes to that end is that of intentionally killing two others.

With the aim of moving forward towards a plausible account of the dignity of persons, I propose we take a step at which Kant would surely balk, namely that of abandoning the notion that the requirement to respect the dignity of persons is a categorical imperative. For Kant "categorical imperative" sometimes refers to an absolutely necessary practical principle: one that all of us, ought always, all things considered, to conform to. Sometimes "categorical imperative" refers to an absolutely necessary practical principle that is also the supreme principle of morality: the principle from which all genuine moral duties derive. Kant's FH is supposed to be a categorical imperative in both of these senses, of course, as is RFH. But on the account I propose, the principle that we ought to respect the dignity of persons is to be understood as a categorical imperative in neither of these senses. I do not intend the principle to be the ultimate source of all moral duties. A person might have a duty, say, not to wantonly cause pain to an animal, which duty would not derive ultimately from a principle commanding that he respect the dignity of persons. And I do not believe that it is never morally permissible, all things considered, to fail to respect someone's dignity. Like the Mere Means Principle, the command to respect the dignity of persons can in my view sometimes be overridden.

Let us call the following the Kant-Inspired Account of Dignity—KID for short.

The dignity of persons is a special status that they possess by virtue of having the capacities constitutive of personhood. This status has several features, including the following:

1. It is such that a person ought not to use another merely as a means. This first aspect of persons' special status is lexically prior to all of those that follow.
2. All persons have a status such that if an agent treats them in some way, then she ought to treat them as having unconditional, transcendent value.
3. The status of persons is such that if an agent treats another in some way, she ought to treat him as having an unconditional, transcendent value that does not change as a result of what the other does or of the agent's relation to him, apart from the following exceptions.
   a) Without violating 3, an agent may treat person B as having a lower value than someone else, A, when A bears certain special relations to the agent, such as being identical to her or being a member of the agent's family, *and* the agent reasonably believes that her treating B in this way is necessary to maintain A's personhood.
   b) Without violating 3, an agent may treat B as having lower value than A if B has used or is using some person merely as a means or B has treated or is treating some person merely as an obstacle. But the agent may treat B as having lower value than A only to the extent that doing so is, according to the agent's reasonable belief, necessary to prevent or curtail B's treating A merely as a means or merely as an obstacle.
4. The status of persons is such that an agent ought to treat them as having a value to be respected, rather than as having a value to be maximized by bringing as many persons as possible into existence.

An agent's treatment of a person respects the dignity of that person only if it accords with the special status just described.

KID obviously requires clarification. From the outset, treating some person in some way needs to be distinguished from treating him as a means or, equivalently, using him. The former is a broader category than the latter. To bring up an earlier example, a police officer who, while in pursuit of a fleeing assailant, pushes aside a bystander blocking her path treats the bystander in some way, but she does not use him. Let us say that someone treats another in some way if she intentionally does something to another in order to promote an end she is pursuing. Sometimes an agent counts as treating another in some way by choosing to pay no attention to her. A pedestrian treats a mendicant in some way if she ignores his request for money in order to promote some end of hers, such as not being distracted from recalling whom she needs to phone when she arrives at work. But she does not thereby use the mendicant. All cases of an agent's using another are also cases of her treating the other in some way, for all of the former involve her intentionally doing something to another in order to promote some end of hers. But, as we have seen, not all cases of an agent's treating another in some way are cases of her using the other.

According to KID (1), persons' dignity is such that agents ought not to use others merely as means. (Conditions under which someone is used merely as a means are specified in the Hybrid, Actual Consent, and Induced Vulnerability Accounts.) The scope of this constraint is intended to extend even to those persons who treat others merely as means and who, according to the account, might without failure to honor their dignity be treated as having a value lower than others. The account includes a constraint against even the journalist treating his two attackers merely as means. (Of course, neither the Hybrid, nor the Actual Consent, nor the Induced Vulnerability Account implies that he is doing this.)

KID specifies that all persons have a status such that if an agent treats a person in some way, then she ought to treat him as having unconditional, transcendent value or worth (2). The concept of unconditional value invoked in KID has much in common with that invoked in the Respect-Expression Approach to FH. According to the concept invoked in KID, something has such value if it has positive value in every possible context in which it exists. There are no conditions, real or possible, under which something that has unconditional value exists but lacks value. For example, if the existence of an animal species has unconditional value, then it has at least some positive value in every context, even one in which there are no sentient beings around who might be aware of its existence. Second, if particular beings possess unconditional value, this value does not vary on the basis of their intelligence or talents alone. In terms of this value as it applies to persons, the most intellectually or physically gifted has no more worth as a direct result of his talent than does the least intellectually or physically gifted. Third, if a particular being has unconditional value, then it has a value that neither varies as a result of its instrumental value to others nor of its impersonal value, that is, the value that an impartial rational spectator would assign to it. A surgeon capable of saving many lives does not thereby have more worth in the relevant sense than a playwright who could save no one, even if an impartial rational spectator would, if she had to choose, prefer the continued existence of the former over that of the latter. Fourth, if a particular being has unconditional value, then its value does not increase or decrease based solely on its level of its health, personal satisfaction (i.e., happiness, in one sense of the term) or well-being. Someone who has contracted a disease that has left him paraplegic and depressed has not thereby lost any of his worth as a person. However, the notion of unconditional value embodied in KID is weaker than that invoked in the orthodox Kantian account. KID does *not* embrace the idea that if particular beings have unconditional value, then this value varies not at all among them. For example, according to KID, it is consistent with holding that a particular being has unconditional value to hold that its value might diminish, though not disappear, depending on the actions it performs.

To say that a being of a particular kind has transcendent value is to say that no amount of anything that is not a being of that kind can have a value equal to a being of that kind. To say that persons have transcendent value is to say that no amount of anything that is not a person can equal the value of a person. It is to imply that persons have value that transcends that of non-persons.

We can, of course, contrast unconditional, transcendent value with mere price. Let us stipulate that if something has mere price, then, first, its having any value depends on someone's being willing to give something for it or at least wanting it to exist and, second, that we can imagine circumstances in which it exists but no one would be willing to give anything for it or even want it to exist. Something of mere price is, by our account, not unconditionally valuable. So according to KID, all persons have a status such that an agent ought to treat them as having value beyond mere price.

KID is *not* committed to the view that persons alone have a status such that they ought to be treated as having value beyond price. The account leaves open the possibility that some non-persons have this status. Moreover, it leaves open the possibility that some non-persons have status such that their use is subject to moral constraints. It would be consistent with KID (but not required by it) to maintain that human fetuses have a status such that they ought to be treated as having a value beyond price. If they have that status, then it would presumably be wrong for a woman to have an abortion simply in order to be able to take a trip to Europe. It would also be consistent with KID (but not required by it) to hold that fetuses have a status such that they ought not to be used in certain ways. One might, for example, hold that it would be wrong to use a fetus in experiments and thereby prevent it from developing into a person, unless doing so somehow furthered the end of preserving persons. There are, of course, other beings (e.g., animal species) or even states of beings (freedom from excruciating pain) that, according to some, demand being treated as having value beyond price. KID also leaves open the possibility that they have a special status that entails that they ought to be treated as having value beyond price.

An agent treats another person as having unconditional, transcendent value, let us say, if and only if, in the given context, the action she performs is among those that someone might perform if he reasonably believed his action to be (successfully and absolutely) constrained by his holding them to have this value.[7] For example, an agent would not be treating another person as having unconditional, transcendent value if she kills him solely in order to prevent someone else from losing half of his inheritance. This action is not among those that someone might perform if he reasonably believed what he did to be constrained by his valuing persons beyond price. He would not kill another simply to improve the balance sheet of a third party. The third party's balance sheet is obviously not the same thing as his rational nature: a person who is poorer than he otherwise might be is still a person. Although his action may well run afoul of KID through violating the mere means constraint, an agent *would* be treating another as having unconditional, transcendent value if he killed the other ultimately in order to save three people. For someone who reasonably viewed her actions to be limited to those that accord with the notion that persons have such value might do that. Treating persons as having unconditional, transcendent value does not itself commit one to treating them as having a value closed to all aggregation. This aspect of the proposal is part of what distinguishes it from the Respect-Expression Approach to FH. The proposal does not incorporate the notion that the dignity of persons requires us to treat them as having incomparable value.

---

[7] The notion of reasonable belief at work here is the now familiar one of belief favored by the evidence available to the agent, given the information he has, his education, his upbringing, and so forth.

At this point it should be clear that a being's having value beyond price does not entail that it has the dignity of a person, as KID characterizes such dignity. For example, let us assume that cats do not have the capacities constitutive of personhood. Nothing in KID rules out the possibility that cats are nevertheless valuable beyond price, that is, that they are valuable no matter whether any person is willing to give something for them or even wants them to exist. According to KID, the dignity of persons is such that if one treats them in some way, one ought to treat them as having unconditional worth. Treating a being as having unconditional worth involves treating it as having worth that does not increase or diminish solely as a result of its health or well-being. But there is nothing self-contradictory in holding that cats have worth beyond price and yet holding it to be legitimate to treat a cat who has a chronic condition that results in its suffering as having less worth than a cat who has no such condition—for example, by choosing, solely on the grounds of the cats' relative well-being, to save the latter over the former when there is not enough medicine to save both. (However, it *would be* inconsistent to hold both that cats have value beyond price and yet refrain from preserving the life of a cat one could easily save solely on the grounds that no one wants it.) If a being does not have value beyond price, then it does not have unconditional value and so does not have dignity. But a being might have value beyond price and yet nevertheless lack unconditional value and thus lack dignity.

According to KID (3), the status of persons is such that, apart from some specified exceptions, an agent ought to treat others as having an unconditional, transcendent worth that does not change as a result of the agent's relationship to them or what they do (or have done). For example, it would violate KID for a public health official to grant a patient's wish to advance on the waitlist for a kidney transplant on the grounds that she went to high school with the patient's mother.[8] Or suppose that a doctor is able either to save one person's life or the lives of two others, but not all three. It would violate KID for the doctor to save the one solely on the grounds that, unlike the others, he has not been convicted of insider trading. This third tenet of KID is not entailed by the second, which requires that if an agent treats a person in some way, she treat him as having unconditional, transcendent value. For example, our doctor might be treating the two persons convicted of financial crimes as having unconditional value (as that value is specified in KID), just not as much value as the one who has no such convictions.

So far as KID is concerned, an agent's treating a person as having lower value than another by virtue of the agent's relations to him is sometimes consistent with honoring the person's status as a being with dignity. It is consistent with this when the other bears some special relationship to the agent, such as being identical to her or being the agent's daughter *and* the agent's treating the person as having lower value is necessary to secure

---

[8] Of course, a doctor's behaving in these ways might be wrong for reasons in addition to that of her thereby failing to respect the dignity of persons. Her behavior might also violate special obligations she takes on by virtue of assuming the role of physician.

the other's existence (3a). For example, suppose that your daughter and two strangers are in imminent danger of drowning and crying for you to help them. You hear their cries, but you are able to save either only your daughter or only the other two. If you save your daughter straightaway, on the grounds that she is your daughter, then you treat the strangers as having less value than she has. But KID does not entail that in saving your daughter you are failing to respect the dignity of the two strangers. Of course, this feature of KID prompts some difficult questions. What qualifies as a special relationship? Close family relations presumably do, but what about friendships or religious affiliations? I leave these questions unanswered.[9] In any case, it is important to note the limited nature of the permissibility (insofar as KID is concerned) of treating those who bear special relationships to us as having more value than others. KID does not, for example, hold as consistent with honoring a stranger's dignity treating her merely as a means. It would not countenance cheating a stranger in order to enrich one's children.

KID also includes exceptions to its tenet that an agent violates a person's dignity if, by virtue of what the person does, she treats him as having lower value than another (3b). According to one exception, the agent's doing this is consistent with her respecting the person's dignity (insofar as KID specifies this status) if it is reasonable for the agent to believe both that the other is herself treating someone merely as a means and that his treating her as having lower value than some person is necessary to prevent or curtail her treating this person in this way. Suppose that, as the journalist is aware, the security officers who are attacking him are using him to earn a paycheck and treating him merely as a means. According to the proposal, the journalist may, without failure to appreciate the officers' status as persons with dignity, treat them as having a lower value than he himself has, to the extent that he reasonably believes that doing so is necessary to prevent or curtail their treating him merely as a means. If the journalist intentionally kills the officers, he treats them as having a lower value than he does; for he destroys them in order to preserve himself. But if he reasonably believes that his killing them is necessary in order to stop them from treating him merely as a means, then his treating them as having a lower value than he has does not amount to his failing to honor their status as persons, according to KID.

KID contains another exception to its designation that when an agent, by virtue of what a person does, treats him as having a lower value than another, the agent fails to honor his dignity. Insofar as KID is concerned, the agent can treat a person in this way when it is reasonable for her to believe that the person is treating another as a *mere obstacle*. An agent treats another as a mere obstacle when she does not use him, but intentionally does something to him, without his having been able to consent to her doing it, and what she does fails to treat him as having unconditional, transcendent value. This account largely comprises concepts familiar to us from our discussion of the

---

[9] My own view is that it would fail to honor the dignity of two strangers in mortal peril to save another such stranger over them simply on the grounds that the one is a citizen of the same country.

Mere Means Principle. The account incorporates the notions of using another and of possible consent embodied in the Hybrid Account. And we have just discussed what it means to treat someone as having unconditional, transcendent value.

To illustrate the notion of treating another as a mere obstacle, suppose that on their way to attack the journalist, the two former paramilitary officers are stopped by a detective. They try to kill her in order to arrive at the alley in time for their hit on the journalist. The officers are treating the detective as a mere obstacle. They do something to her without having given her an opportunity to avert their doing it by withholding her consent to it. And what they do, namely try to kill her in order to get in position to kill another and thereby get a big paycheck, does not amount to treating her as having unconditional, transcendent value. Their action is not among those that, in the given context, someone might perform if he reasonably believed his action to be (successfully and absolutely) constrained by his holding persons to have this value. According to KID, the detective does not fail to honor the officers' status as persons with dignity if she treats them as if they have a value lower than hers, insofar as she reasonably believes that her doing so is necessary to stop them from treating her as a mere obstacle. In this case, that implies that her intentionally killing the officers in order to save herself might be consistent with respecting their dignity.

But if the detective kills the officers, would that not entail that she fails to treat persons—in particular the officers—as having unconditional value? Despite appearances, it would not entail this. According to our notion of unconditional value, if someone has such value, then he is valuable in every context in which he exists. Moreover, this value does not vary solely as a result of his intelligence, talents, health, well-being, usefulness to others, or even the worth he would be assigned by an impartial rational spectator. An agent treats persons as having unconditional value if, in light of her situation, the action she performs is among those that someone might perform if he reasonably believed his action to be (successfully and absolutely) constrained by his holding them to have such value. The detective's action is among those someone might perform if he reasonably believed his action to be thus constrained. For someone might reasonably believe that the action of killing the security officers is something that someone who views his action as constrained by holding them to have unconditional worth might do. Someone in the detective's position might reasonably think that, although the officers' actions do not deprive them of all such worth, their actions lower their worth enough such that her worth is greater than that of the two of them put together. And since (we are assuming) it is reasonable for her to believe that the only way to save herself is to destroy the two, an appreciation of the value inherent in persons gives her reason to kill them.

In sum, an agent's treatment of some other person respects the dignity of that person only if it accords with the special status that having dignity amounts to. An agent's treatment of another fails to accord with this special status if she treats him merely as a means or as lacking unconditional, transcendent value. Moreover, an agent's treatment of a person clashes with this special status if, as a result of what the person does or of his relations to her, she treats him as having a lower unconditional, transcendent worth than other persons, apart from some exceptions we have just discussed. It bears repeating that KID is not intended to be a complete account of the dignity of persons. KID leaves open the possibility that, according to reflective common sense, there are ways of violating a person's dignity in addition to those it identifies.[10]

5.5 Applying KID

It is already apparent that the journalist portrayed in Chapter 2 as being attacked by one officer does not disrespect his dignity, according to KID. Nor does the journalist portrayed here disrespect the dignity of the two officers he tries to kill. In neither case is he treating anyone merely as a means. Although in both cases he treats his attackers as having less value than he does, this treatment is consistent with his honoring their dignity so far as KID is concerned, since he believes reasonably that it is necessary to prevent them from treating him merely as a means. That the account has this result supports the idea that it is more plausible than the notion of dignity we explored in Chapter 2.

But how does the account fare regarding the other kinds of examples we examined there? Consider first cases of self-sacrifice such as that of a soldier who allows himself to be killed by a grenade in order to save four others. The soldier treats no other merely as a means. Moreover, his action treats persons as having unconditional, transcendent value. He allows himself to be killed, but he does so in order to preserve other persons. He does not exchange himself for something of mere price, such as an armored vehicle. And, again, treating something as having transcendent, unconditional value is consistent with treating it as having less value than a greater number of beings who also have such value. Finally, the soldier does not treat anyone as having a value that is lower than

---

[10] One might, of course, take a different direction in reconstructing a Kantian account of dignity than the one I have taken in developing KID. Thomas Hill, Jr., suggests, for example, that holding persons to have dignity amounts to holding there to be a set of prescriptions that we are rationally compelled to use in deliberations regarding policies on permissible or impermissible ways of behaving towards them ("Treating Criminals as Ends in Themselves" *Jahrbuch für Recht und Ethik* 11 (2003)). The prescriptions are higher level principles that govern decisions regarding which lower level principles to adopt. Hill suggests that these prescriptions include, but are not limited to, the following (25): First, we "must treat persons only in ways that we could in principle justify to them as well as to all other rational persons who take an appropriately impartial perspective." Second, the value of persons is not commensurable, so we must not compare or weigh the value of one against that of others. And, third, persons have a moral status that limits how we may use them, even to realize good ends. The normative implications of these prescriptions are far from obvious. But they seem to be problematic in some of the cases we have examined. Consider a principle that would permit, but not require, a person to voluntarily kill himself (e.g., by jumping on a grenade) in order to maximize the preservation of persons. It seems as if this principle would run afoul of the second prescription; for it treats the value of persons as commensurable. Since, in my view, the principle in view is plausible, Hill's second prescription has questionable implications. Many other cases illustrate problems with this prescription, I believe. It brings with it difficulties analogous to those an orthodox Kantian account faces as a result of its holding persons to have incomparable worth. These difficulties come into view at several points in the book (e.g., 2.4, 2.5, 6.2, and 6.3). But the rational prescription approach might be developed in a way that would enhance its status as a rival to KID.

that of any other person, so that part of KID does not apply. KID avoids the implication that the soldier's action dishonors the dignity of persons. This result is much more plausible than that generated by the Respect-Expression Approach to FH.

However, in the case discussed earlier (2.3) of a physician's granting a patient's request to be taken off life support, KID has the same implications as the Respect-Expression Approach. In that case, an ALS patient's severe, irreversible disability causes him intense suffering. Even though his rational nature is fully intact, he wants his doctor to disconnect him from a respirator. The physician disconnects the patient, intending him to die so that his suffering will cease. According to our account, the physician is using the patient. She is intentionally doing something to him, namely stopping his respiration, so that his suffering will come to a halt, and she intends his body to contribute (in effect, by shutting down) to her realizing this end. But the physician is obviously not treating her patient merely as a means. The patient can avert the doctor's use of him by withholding his agreement to it. If he tells the doctor not to take him off the respirator, then she will refrain from doing so. However, in disconnecting the patient from the respirator, the doctor is not treating the patient as having unconditional, transcendent value. It would not be reasonable for her to believe that, in the given context, disconnecting him is among the actions that someone who places unconditional, transcendent value on persons might perform, if she saw her action constrained by her placing this value on them. According to KID, humanity has a status such that it ought to be treated as being not only valuable in every context, but more valuable than anything other than humanity, including freedom from intense suffering. (That does not mean that freedom from suffering is unimportant, of course. It might even be unconditionally valuable.) The doctor would be exchanging the patient's rational nature for the cessation of his suffering; and it is not reasonable to believe that that is something someone would do who sees his action as constrained by holding rational nature to have unconditional, transcendent value. So the doctor would fail to respect the patient's dignity if, in accordance with his request, she took him off the respirator so that he would die and cease to be in pain.

Some might find this implication implausible. But the following observations might mitigate their dissatisfaction. As described in Chapter 2, the patient sets and pursues a variety of ends; he is able to engage in projects other than that of eliminating his suffering. So it is not the case that his rational nature is, as it were, permanently exhausted by his striving to be in pain no more. If it were permanently exhausted, then KID would presumably not imply that the doctor was failing to respect the patient's dignity. For it would be reasonable for her to believe that her actions were among those someone might perform if she believed herself to be constrained by holding persons to have unconditional, transcendent value. In these circumstances, it would be reasonable to think that the usual distinction between the capacity of rational choice and a particular exercise of this capacity fails to obtain. In respecting the agent's exercise of her capacity of rational choice, that is, her striving for the cessation of her suffering, one would be respecting all that was left of the capacity itself. One would, in effect, be honoring something of unconditional, transcendent value by bringing about its destruction.

A second observation is that many of us would not wish to embrace the conclusion that, according to a common sense notion of dignity, aiding a competent person to end his life is *always* respectful of his dignity. In Chapter 2 (2.3), we discussed the case of a man who is bent on attracting attention to his poetry. He reasonably believes that the only way to get a significant number of people to engage with his work is for him to die a violent death by another's hand. Let us assume that he is nevertheless capable of pursuing other ends; his capacity of rational choice is not employed solely in his attempt to achieve this aim. It would be implausible to say that we would be respecting his dignity by granting his request to bring about his death. But if we hold on to the view that it would be respectful of the dignity of the ALS patient (according to a common sense notion of it) to do what is necessary for him to attain the end of no longer suffering, then how can we consistently deny that it would be respectful of the poet's dignity to do what is necessary for him to attain the end of getting his poetry noticed? Is it because the poet's suffering at the lack of attention to his work is less severe than the patient's suffering from his disabilities? What if the poet's suffering is very severe indeed? There might be a price in plausibility to be paid by insisting that it is disrespectful of the patient's dignity for the doctor to take him off the respirator. But there might also be a price in plausibility to be paid by insisting that it is not disrespectful of it for her to do that. For then one is forced to construct a well-grounded distinction between this case and others that in some respects resemble it, but in which it seems incorrect to conclude that the deadly pursuit of another's aims respects his dignity.

Another case it might be helpful to reexamine in light of KID is that of Trolley. We have already found that if you operate the switch and direct the trolley onto the track where it will kill one person, you are not treating that person merely as a means. Now we can add that you are treating everyone involved as if he has unconditional, transcendent value. It is true that, as you are aware, your action results in one person's death. But, as you are also aware, it will result in the preservation of five people who would otherwise perish. It is reasonable for you to believe that maximizing the preservation of persons is the sort of thing that a bystander who holds his actions to be constrained by his view that they have such value might do.

But what if you choose not to operate the switch and, as you were aware would happen, the train careens into the five killing them but leaving the one unharmed? Do you thereby fail to respect the dignity of the five? I believe that the answer is no. For among the things someone might do if he reasonably believed his treatment of others to be constrained by his placing unconditional, transcendent value on persons is to refrain from operating the switch. That one's treatment of others is constrained by his valuing persons in this way is consistent with its being constrained by his valuing them in other ways as well. There is at least one way of extending the notion that persons have unconditional, transcendent value such that if one's treatment of others was constrained by this extended notion, one would not deploy the switch. I am thinking of the Kantian notion that persons have unconditional *and incomparable* value. Holding them to have such value entails that they have unconditional, transcendent value, for it entails that they have an unconditional worth that cannot legitimately be exchanged for anything.[11] So, rationally speaking, if an agent's treatment of another is constrained by her holding persons to have unconditional and incomparable value, it is also constrained by the notion that they have unconditional and transcendent value. Moreover, it would be reasonable for an agent who viewed her treatment of another to be constrained by her holding persons to have unconditional and incomparable value to refrain from deploying the switch. She might refrain from doing so on the grounds that deploying it would amount to her treating persons as having comparable value. For her basis for deploying the switch would be the idea that five people are more valuable than one. KID does not imply that failing to maximize life-saving in trolley type cases always dishonors the dignity of persons.

A final group of examples it is helpful to consider here focus on KID's implications regarding taking risks that threaten personhood. To begin, suppose that a government plans to blast a tunnel through mountains. The new tunnel would cut in half travel time between two major cities. The construction project would employ hundreds of workers for two years, but some of them would be doing dangerous jobs. The government's actuaries have advised it that, in all probability, between three and ten workers would lose their lives in the tunnel's construction. If the government goes ahead with the project, it will not treat any of its workers merely as means. Its policies have not foreseeably put any potential workers in a position such that in order not to suffer a significant decrease in well-being, they must sign on to the dangerous work. And the government makes sure that all potential workers are informed of the serious risks they would face on the jobsite. An obvious issue to consider is whether the government, if it went ahead with the tunnel project, would be failing to respect the dignity of its workers by failing to treat them as having unconditional, transcendent value.

Might someone in this context who reasonably believed his treatment of the workers to be constrained by his valuing persons in this way use them to build the tunnel? At first the answer might seem to be obviously no. For reducing travel time between cities is a matter of convenience, and an increase in convenience presumably has mere price. But things would not be this simple. Cutting down travel time might also be associated with reducing traffic accidents, including fatal ones, as well as with significantly reducing air pollution and thereby reducing morbidity and mortality from respiratory disease. Depending on the facts on the ground, someone might reasonably believe that more persons would be preserved if the tunnel got built than if it did not.

---

[11] The fact that the orthodox Kantian notion of unconditional value differs from the one embedded in KID makes no difference here, for if something is unconditionally valuable according to the former, then it is also unconditionally valuable according to the latter.

So, depending on the facts on the ground (which, of course, might be difficult to assess), the government's treatment of its workers might respect persons' dignity.

There are, of course, situations in which someone's taking a mortal risk would not honor his dignity. Someone who believed his treatment of persons to be constrained by their having unconditional, transcendent value would not go snow-boarding in an avalanche zone solely for the thrill of it. So if a practitioner of extreme sports did that, she would not be honoring her dignity.

But then does honoring our dignity require us to minimize absolutely risk to our lives, apart from risk involved in attempts to preserve the lives of others? Driving in an automobile carries a risk of death. In 2008, there were 37,261 motor vehicle crash fatalities in the United States.[12] Does the proposed account require us to drive only the safest of cars, engirdled with roll bars and bedecked with flashing yellow lights, or to give up driving altogether, assuming that such behavior would in fact decrease our risk of premature death?

I believe that the answer is no and that considering a case that does not concern risk to persons will help us to see that. Suppose that a museum director holds a sculpture to have extraordinary aesthetic value. In order to keep risk of the object's losing this value (e.g., by being destroyed) to an absolute minimum, she would have to keep it locked in an environmentally controlled vault, deep underground, in a remote location. But reasonably believing her action to be constrained by her valuing the sculpture in this way, she might also subject it to greater risk, by, for example, allowing it to be displayed in her museum or even by loaning it out to a travelling exhibition, many of us believe. We believe this, I suspect, because we hold something like the following. Part of honoring high aesthetic value is to promote its appreciation by many people. And the director's allowing the sculpture to be in less than maximally safe circumstances does just that. There is, of course, a balance between honoring the value of an art object through trying to ensure its continuing existence and honoring it by promoting its appreciation. Conditions (e.g., war, civil unrest, or even poor air quality) might sometimes be such that taking the risk involved in doing the latter fails to harmonize with holding the art object to have extraordinary aesthetic value. But whether conditions are like this will be a matter of judgment in particular cases.

In much the same way that respecting an object's extraordinary aesthetic value can involve doing something other than minimizing the risk that the object will be destroyed, so respecting a person as having unconditional, transcendent value can involve doing something other than minimizing the risk that he will lose his life. What gives the person status to be treated as having unconditional, transcendent value, according to KID, is his having a certain set of capacities, including those to set and rationally pursue ends. But part of honoring a person as having such value, it seems, is

---

[12] Anders Longthorne, Rajesh Subramanian, and Chou-Lin Chen, "An Analysis of the Significant Decline in Motor Vehicle Traffic Crashes in 2008," US Department of Transportation (Washington D.C.: National Highway Traffic Safety Administration, 2010), 1.

sometimes honoring the *exercise of these capacities*, including his actually setting and rationally pursuing ends. If one did not sometimes honor the exercise of the capacities that constitute personhood, then why would one honor the capacities themselves? If we did not under some circumstances honor acquiring scientific knowledge, for example, why would we honor the capacity to acquire it? Or, hitting closer to Kantian home, if we never honored good willing (i.e., very roughly, doing something because one believes it to be required by moral law), why would we honor having a good will (i.e., very roughly, the disposition to do things because one believes them to be required by moral law)?

We need to proceed carefully. It does not seem self-contradictory to honor a capacity, but never honor its exercise. For example, suppose that a theologian holds that one would not have the capacity to do good unless one also had the capacity to do evil. We can imagine her insisting without self-contradiction that the capacity to do evil is worthy of honor, but that no particular exercise of the capacity is worthy of it. But note that, according to the theologian, there is a basis for honoring the capacity to do evil that is never a basis for honoring a particular exercise of this capacity, namely the capacity's role in making possible the doing of good. There does not, however, seem to be a basis for honoring the capacities that constitute personhood as having unconditional, transcendent value that does not also constitute a basis for honoring some exercise of these capacities.

In any case, the main point here is that, in order to honor himself as having unconditional, transcendent value, a person need not minimize the risk to his person. He can sometimes honor his personhood by exercising the capacities constitutive of it. Driving a car and engaging in similarly risky activities are obviously often part and parcel of the rational pursuit of ends. Such actions are among those that might be performed by someone who placed unconditional and transcendent value on himself and reasonably saw his actions to be absolutely and successfully constrained by his doing so. But honoring one's rational nature involves striking a balance in one's actions between securing the continued existence of one's rational nature and exercising the capacities constitutive of this nature. The snow boarder described above presumably fails to strike such a balance, as would someone living in a tranquil town who "for safety reasons" never left his house. But no more in the case of honoring persons than in that of honoring art objects does it seem possible to specify precisely where that balance lies.

5.6 Honoring Dignity as a Defeasible Constraint

According to KID, an agent's treatment of another fails to respect his dignity if he treats the other merely as a means. I claimed earlier that it is, all things considered, sometimes morally permissible to treat another merely as a means. So I am committed to the view that it is, all things considered, sometimes morally permissible to fail to respect the dignity of persons. Failing to respect the dignity of persons is, in my view, always wrong *pro tanto*, but not necessarily wrong on the whole. But when is it morally permissible to fail to respect the dignity of persons? I do not have a full or fully precise answer to this question, and I doubt whether it admits of such. However, in Chapter 3 I introduced a Wrong-Making Insufficiency Principle, which attempts to specify one set of circumstances in which one's *pro tanto* reason not to treat others merely as means does not amount to an all things considered reason not to do so. With the help of concepts introduced in conjunction with KID, I here develop a second principle of this sort. I then offer a few further ideas about when, according to considered views held by many of us, failure to respect the dignity of persons might not be wrong, all things considered.[13]

The Wrong-Making Insufficiency Principle (3.5) holds roughly that an agent's treating another merely as a means is generally not wrong, all things considered, when in so doing she is pursuing an end that the other cannot share as a result of his treating someone merely as a means. An example of such a case is that of Attack, in which a congregant uses one person attacking his church to buy time for fellow congregants to escape. The congregant treats the attacker merely as a means. But it is reasonable for the congregant to believe that what, rationally speaking, prevents the attacker from sharing his end is (roughly) that the attacker is treating congregants merely as means.

We are now in position to set out a second principle specifying conditions in which an agent's treating another merely as a means does not typically amount to her acting wrongly. This principle holds that an agent's treating another merely as a means is generally not wrong, all things considered, when in so doing she is pursuing an end that the other cannot share as a result of his treating someone merely as an obstacle. More precisely, the Second Wrong-Making Insufficiency Principle states:

Suppose that an agent, A, treats another, B, merely as a means. A's doing this does not typically suffice to render her action morally impermissible, if the following obtains: It is reasonable for A to believe that B is, rationally speaking, prevented from sharing the end A is pursuing in using B as a result of an end B is pursuing in the course of treating someone as a mere obstacle.[14]

---

[13] One might claim instead that failing to respect the dignity of a person *is always* morally wrong, but that an action's being morally wrong does *not* always generate a rational requirement to refrain from performing it. In other words, while we are always morally bound to respect the dignity of persons, we are not always rationally bound to do so. I do not find this claim plausible. If, in treating someone merely as a means, an agent kills him and thereby takes the only route available to her to saving the lives of a million people, then the agent is simply not acting wrongly, in my view. Moreover, the police officer who treats the white-supremacist merely as a means to getting information about a planned attack is not acting wrongly, nor is the congregant who uses one attacker merely as a means to foil another (3.5).

[14] For the sake of ease of expression, I have simplified the condition specified in this principle. Strictly speaking, it should read: Suppose that an agent, A, treats another, B, merely as a means. A's doing this does not typically suffice to render her action morally impermissible, if the following obtains: It is reasonable for A to believe that B is, rationally speaking, prevented from sharing the end A is pursuing in using B as a result of an end B is pursuing (*or has pursued or will pursue*) in the course of treating someone as a mere obstacle.

A person treats another as a mere obstacle when she does not use him, but intentionally does something to him, without his having been able to consent to her doing it, and what she does fails to treat him as having unconditional, transcendent value. An illustration of this principle is easy to sketch. Suppose that a religious relic has been housed for hundreds of years in a mosque. A group of thugs has broken in to destroy it. To get to the relic, they shoot the guards in the way. If one of the guards treats an attacker merely as a means, say by pushing him into one of the other attackers so that he can retreat to a defensible position, he is acting wrongly *pro tanto*. But the conditions specified in the Second Wrong-Making Insufficiency Principle are fulfilled. It is reasonable for the guard to believe that his attackers are treating him as a mere obstacle. Of course, they have given him no opportunity to avert their attack on him by withholding his consent to it. And by shooting him in order to be able to destroy a relic, they are obviously not treating him as having unconditional, transcendent value. Moreover, it is reasonable for the guard to believe that the attacker whom he treats merely as a means is, rationally speaking, prevented from sharing his end of gaining a defensible position as a result of his pursuing the end of destroying the relic. That the conditions specified in our principle are realized helps explain why we do not hold the guard's using the attacker merely as a means to be wrong, all things considered.

Our two wrong-making insufficiency principles set out relatively specific conditions under which failure to respect a person's dignity tends not to be wrong, all things considered. But we can also make some general conjectures as to when such a failure can be morally permissible. As stated in Chapter 3, I subscribe to the view, which, I believe, is shared by many, that it can, all things considered, be morally permissible to treat another merely as a means when doing so is the only way to save a great number of lives. It can be morally permissible to treat someone merely as a means by deceiving him so that he unwittingly reveals information that leads to the saving of thousands of innocent people, for example. It can also in my view be morally permissible to treat one person merely as a means and in so doing foresee that one will severely injure or even kill him, if treating him in this way is the only way to reveal information that leads to the saving of thousands of innocent people.

But are there situations in which it is morally permissible to treat someone merely as a means and foresee that one will thereby kill him, for an end that does not involve preserving persons? Perhaps it is sometimes morally permissible to do this if it is necessary to preserve some animal species that includes no persons among its members. Or perhaps it is sometimes morally permissible to do so it if it is necessary to reduce horrible pain among many persons.[15] But note that if the pain of the many is intense enough, then its reduction in them might really amount to a preservation of person-

---

[15] Science fiction might help here. Imagine that one person's body and it alone contains some compound that can be synthesized into a medicine that would quell the intense pain of many persons. The only way to gain access to the compound is to kill the one, but she is unwilling to die to prevent others' pain.

hood. For at its highest intensity pain can cause people to lose themselves, that is, to be incapable of doing anything but endure it.

These reflections support the following conjecture regarding the considered moral judgments many of us embrace, namely those of us who hold that persons and some other things have value beyond price. Among the cases in which it is, all things considered, morally permissible to treat a person merely as a means are cases in which one reasonably believes that doing so is necessary to secure the existence of a person or some other thing that has value beyond price. Moreover, it is not (or at least is rarely) the case that it is, all things considered, morally permissible to treat a person merely as a means if one *does not* reasonably believe that doing so is necessary to secure the existence of something that has value beyond price.

Thus far we have focused on the moral permissibility of treating another merely as a means and thereby failing to respect another's dignity. But is it ever morally permissible to fail to respect another's dignity when one is not treating the other merely as a means?

Before addressing this question, we need to explore a bit more what the implications for our actions of respecting persons' dignity would be. In particular, it will be helpful to investigate what respecting persons' dignity would require in terms of life-saving aid to strangers. KID ranges over our *treatment* of others. A person treats another in some way if she intentionally does something (or refrains from doing something) to another in order to promote an end she is pursuing. Suppose that, as I realize, the only way a stranger can preserve his life is to receive financial aid from me. If the stranger requests aid from me and I deny it on the grounds that granting it would prevent me from buying the new car I'm set on, then I treat the stranger in some way. Do I also fail to respect his dignity? (For the sake of simplicity, let us suppose that, as I am aware, my granting his request would have no practically relevant effects besides his continuing to live and my not being able to buy a new car. And the effects of my denying his request would be similarly limited.) I am not using the stranger: I do not intend his presence or participation to contribute to my buying the new car. So I am not treating him merely as a means. But am I treating him as having unconditional, transcendent value? Is my denying the stranger's request so that I can buy a new car among the actions that someone might perform if he reasonably believed his action to be (successfully and absolutely) constrained by his holding persons to have such value?

I think that the answer is no. It would not be reasonable to believe that if one's actions were constrained in this way, one would in the context described choose to obtain a car (i.e., something of mere price) over preserving a person (i.e., something that has unconditional, transcendent value). Acting with respect for the special value of a thing can and often does involve trying to preserve that thing. Suppose, for example, that we hold a certain painting to have exceptional aesthetic value. One way of respecting this value is to try to maintain the painting in existence by, say, protecting it against destruction from insects, excessive heat, and so forth. Or suppose that we hold a stand of 1,000-year-old Sequoia trees to be of special worth. One way of respecting this worth would be to do what is in our power to prevent the forest from being destroyed to make way for a mall. If we hold our treatment of a person to be successfully and absolutely constrained by our view that he has unconditional, transcendent value, then it seems that our treatment of him would, if practically possible, involve an attempt to preserve him if his life (and thereby his personhood) was at risk. In some special cases, for example, if the person's rational nature is irrevocably exhausted by a will to die, then our treatment of him might not involve an attempt to preserve the person. But here we are not addressing such cases.

If our reasoning thus far is sound, then we can draw some conclusions regarding the implications of KID. Suppose that wealthy people can provide life-saving aid to distant poor people who, without it, would die prematurely. The wealthy people are aware of this, and aware as well that a way to promote the saving of lives, not just in the short run or among those distant poor people, but overall, is to provide such aid. The poor request life-saving aid from the wealthy. But in order not to be derailed in pursuing ends that, the wealthy acknowledge, fall short of having unconditional, transcendent value, the wealthy decline their request. Ends that fall short of having such value need not be "materialistic" in an everyday sense, of course. Among such ends might be some that are arguably beyond price such as that of freeing a terminally ill child from excruciating pain as his life comes to an end. Other ends might not be beyond price, but nevertheless contribute to one's leading a flourishing life. An example might be the end of learning to play the violin. Nevertheless, according to KID, even in pursuing such ends rather than preserving the lives of the poor, the wealthy would be failing to respect their dignity.

In my view, this implication casts doubt on the position that we have a categorical obligation to respect persons' dignity. I think it makes sense to say that in choosing to pursue an end that we have no reason to treat as having unconditional, transcendent value rather than to preserve someone in his personhood, we fail to respect that person's dignity. But our pursuit of such an end can be morally permissible, all things considered. A second conjecture is that many of us who believe that some things are without price judge the following. Actions that fail to respect the dignity of persons but that do not involve using anyone merely as a means are often morally permissible, all things considered, even if in performing them we do not aim to secure anything we have reason to treat as having value beyond price, but rather something we take to be important for the well-being of ourselves or those about whom we care most. An example of such an action might be that of using money that could fulfill a psychologically and personally distant person's appeal for life-saving financial aid for a child's violin lessons instead.

A third and final conjecture concerns conditions under which an action is, all things considered, morally permissible, even though, according to KID, it fails to respect the dignity of some person. Take two actions, both of which fail to respect the dignity of a person. One action is done in a context in which one has been using this person, while the other action is not. Other things being equal, we are more open to judging the latter than the former action to be morally permissible, according to the conjecture.

The strength of the reasons required to override our *pro tanto* reasons to respect the dignity of a person tends to be greater if we are using the person than if we are not using him.

Suppose, for example, that the distant stranger who appeals to you for life-saving financial aid is someone you are using, with her informed, voluntary consent, to do computer programming for you. You do not know her personally; the two of you have had only brief, professional correspondence. It seems that you would need stronger reason to be justified, all things considered, in ignoring her request for life-saving financial aid than to be justified in ignoring the request of someone whom you were not using. We would be less open to concluding that it was morally permissible for you to give your child violin lessons instead of saving the former person than it would be to give her lessons instead of saving the latter.

The notion that our using a person increases the strength of reasons required to override our *pro tanto* reasons to respect his dignity does not seem to be significantly or, perhaps, even at all diminished in cases where not only we ourselves, but also the person we are using benefits from the use. Suppose, for example, that you pay the distant programmer more than what she could get from a local employer. Other things being equal, you would, many of us judge, nevertheless need stronger reasons to render justified your failing to respect her dignity than you would to justify your failing to respect the dignity of someone you had never used at all.

In sum, we have isolated two principles that specify conditions under which an action is typically morally permissible, all things considered, even when it involves a failure to honor someone's dignity. These Wrong-Making Insufficiency Principles imply roughly that an agent's using another merely as a means is generally not wrong when she is doing so in the course of realizing an end that the other cannot share either as a result of his (the other's) treating someone merely as a means or of his treating someone merely as an obstacle. Moreover, we have made three general conjectures regarding when we judge a failure to honor someone's dignity to be morally permissible, all things considered. First, suppose an agent treats another merely as a means. Among the cases in which her doing so is morally permissible, all things considered, are ones in which she reasonably believes that it is necessary to secure the existence of a person or some other being that has value beyond price. But it is rarely the case that her just using another is, in the end, morally permissible if she does not reasonably believe it to be necessary to secure the existence of something with such value. Second, suppose that in performing a certain action, an agent is not treating another merely as a means. If, in performing it, she nevertheless fails to honor someone's dignity, her action is sometimes morally permissible, all things considered, if she does it in order to secure something important to her well-being or to the well-being of those she cares about most. Third, suppose again that an agent is not treating another merely as a means. If she is nevertheless using the other, the *pro tanto* wrongness of her failure to honor the other's dignity is, other things being equal, more likely to amount to all things considered wrongness than it would be if she were not using him.

In the end, one might reject the notion of respecting the dignity of persons contained in KID. But if one does, it should not be because KID is hopelessly vague, but rather because some other moral view squares better than it does with one's considered moral judgments (regarding principles and cases) and background theories (e.g., regarding human psychology). Like FH on the Respect-Expression Approach, KID is determinate enough to be evaluated as an option in normative ethics.

Deflationists hold that the idea that persons' dignity ought to be respected adds nothing ethically significant to a familiar idea that persons ought to be respected (Chapter 1). According to the latter notion, we ought not to interfere with the choices of a competent, well-informed person, unless those choices harm another. In medical contexts, we must get persons' voluntary, informed consent before treating them or using them in research, as well as protect their confidentiality. The chapters that follow will (among other things) help to illustrate that one might respect a person according to the familiar idea, and yet fail to honor his dignity, according to KID. If we judge such failure to be morally significant, as I believe we do, then a notion of dignity can play an important role in the moral evaluation of how we treat persons.

PART II

Practice

6

Allocation of Scarce, Life-Saving Resources

In the distribution of resources, persons must be respected, or so many philosophers contend.[1] Unfortunately, philosophers often leave it unclear why a certain allocation would respect persons, while another would not. In this chapter, I explore what it means to respect persons in resource allocation—specifically in contexts in which scarce, life-saving resources must be distributed.

As a way of grounding discussion, let us focus on two sorts of cases. We assume in both that our task is to allocate a life-saving resource between different persons. In helping these persons (or a subset of them), we are not discharging a duty of beneficence. Each person needs and wants to get the resource. But since the resource is scarce, we cannot make it available to all. Each person has a claim on the resource in the relatively weak sense that it would be wrong for us to refrain from giving it to her on morally arbitrary grounds (e.g., because we do not like her) or on grounds inappropriate to the context (e.g., because she is not a close friend). Finally, no person in our cases is morally responsible for her need of the resource in any way that would affect her claim on it.

In the *different-age case*, we have one indivisible life-saving drug and two patients who are identical in every relevant respect except that one of them is 20 years old and the other one is 70 years old. The patient who does not get the drug will die. If the younger person gets the drug, she will live for many years yet; if the older person gets the drug, she will die of natural causes in a few years.

Many people believe that the drug should be given to the 20-year-old patient. One consideration in favor of this choice is that giving the drug to the younger patient does more good: since she will live longer, saving her life creates a larger benefit. Another consideration that many find relevant is that the older person has had a longer life: she has already had her "fair innings." Perhaps it would be unfair to the younger person to

[1] See, for example, John Harris, "QALYfying the Value of Life," *Journal of Medical Ethics* 13 (1987), 120, and Martha Nussbaum *Frontiers of Justice: Disability, Nationality, Species Membership* (Cambridge, Mass.: Harvard University Press, 2006), 92.

deny her the chance of a full human life, given that the older person has already had such a life.[2]

It is not clear, however, that giving priority to the younger person is compatible with respecting persons—at least if doing that amounts to treating the two patients with the equal respect that is morally owed to them. Assuming that both patients want to be saved and they both have a claim on the drug, it seems that we show *less* respect for the older person since we respect her wish to go on living *less* than the wish of the younger person. Giving priority to the 20-year-old seems to involve *not* treating both patients with equal respect, although it seems to be the recommended course of action on both benefit-maximizing and fairness grounds.

In the *different-number case*, we have to decide whether to save one person or five persons from certain death. Perhaps there were two traffic accidents, and one person was injured in the first and five persons were injured in the second. All the accident victims have immediately life-threatening injuries, and they are alike in all relevant respects. We can, however, reach only one of the accident scenes. If we save the one person, the five persons in the other accident will die; if we save the five, the one person will die.

On benefit-maximizing grounds, we should save the five. But some philosophers argue that it would be unfair to give no chance to the one person at all: if we did that, we would fail to show proper respect for her—perhaps because her claim is not taken into consideration at all. On this view, benefit-maximizing and fairness considerations point in different directions. Others argue that it is possible to save the five persons while giving no chance to the one person without failing to show proper respect for her—perhaps because her claim is taken into consideration just as much as those of the others, but it is outweighed by them in one way or another. In this example, both sides appeal to an idea of respect for persons.

The chapter uses these examples as test cases against which the implications of accounts of respect for persons can be assessed. The cases are stylized, but nonetheless relevant to actual ones. Choices about life-saving must be made in the allocation of organs for transplant, vaccination and treatment in pandemic flu, and beds in intensive care units. In resource-poor countries, treatment priority-setting (e.g., deciding who does or does not receive antiretroviral therapy for HIV infection) is an unfortunate necessity.

Two of the accounts of respect for persons that the chapter applies to our cases (6.2 and 6.3) are familiar: the Respect-Expression Approach to Kant's Formula of Humanity (FH) and the Kant-Inspired Account of Dignity (KID). On both of these

---

[2] For the fair innings argument, see John Harris, *The Value of Life: An Introduction to Medical Ethics* (London: Routledge & Kegan Paul, 1985); for a recent discussion, see Greg Bognar, "Age-Weighting," *Economics and Philosophy* 24 (2008).

accounts, respecting persons amounts to respecting their dignity. (Of course, the accounts embody different conceptions of dignity.)

But why consider further implications of the Respect-Expression Approach, given that we have already put the plausibility of its principle (RFH) into doubt? The finding that RFH has implausible normative implications in the kinds of cases we examined in Chapter 2 significantly damages its credibility. But the finding does not itself constitute a sufficient basis for concluding that we are unjustified in embracing it. In order to establish that conclusion, we would (at least) have to look at RFH's implications in a wider range of cases, as well as to compare the plausibility of its implications with those of rival principles, such as the ones sketched in Part I. The criticism of RFH was not intended to amount to a definitive refutation of it, but rather to an exposure of weakness sufficient to warrant the serious exploration of alternative Kantian principles. The process of reaching reflective equilibrium is ongoing.

In any case, the chapter also explores the implications regarding our cases of a third account of respect for persons (6.1), one that places a special value on persons by virtue of their having a set of psychological capacities, including autonomy. The chapter argues that KID has an important advantage over the other two accounts—an advantage that underscores the value of exploring KID as a rival to other accounts of the worth or dignity of persons. The chapter ends by contrasting KID with a particular benefit-maximizing view of the distribution of scarce resources, namely one that uses the Quality Adjusted Life-Year as a measure of improvement in health-related quality of life (6.4).

Respecting the dignity of persons is sometimes understood simply as acting toward them in ways that are morally justified. If it is justified to give the life-saving drug to the younger person, for example, then doing so respects the dignity of both the younger and the older. On this view, to say that the dignity of persons has been respected is just a shorthand way of saying that some action or policy is right, permissible, or just. But here as elsewhere in the book, I consider the demand that the dignity of persons be respected to have its own moral weight. It makes sense to say that an action, policy, or institution is wrong *because* it fails to honor the dignity of persons.

6.1 The Equal Worth Account

One familiar account, suggested, for example, by Jeff McMahan, connects the notion of respect for persons to the idea of the worth of a person.[3] On this view, persons have a special value, or worth, in virtue of having certain psychological capacities: the capacity to perceive and understand the world, to form desires and life plans, to reflect and deliberate on these desires and life plans, and so on. Chief among these capacities is the person's autonomy or "the capacity to direct one's life in accordance with values that

---

[3] Jeff McMahan, *The Ethics of Killing: Problems at the Margins of Life* (New York: Oxford University Press, 2002), 256, 261, 478.

one reflectively endorses."[4] This notion of autonomy is, of course, of the same sort as the split-level notion introduced in Chapter 1 (1.3). We shall call this view the Equal Worth Account.

A person's worth is a threshold concept: each person who has autonomy (as well as the other necessary psychological capacities) has worth equal to that of every other person, no matter how well- or ill-developed her autonomy (or those other capacities) may be. Those whose autonomy or other cognitive capacities fall below the threshold do not count as persons in the full moral sense, and hence have lower or no moral worth.

The worth of a person is entirely independent of how well her life goes. Thus, this sort of value does not vary according to the level of a person's well-being. Moreover, a person's worth is neither a function of her instrumental value to others nor of her impersonal value, that is, of her value from the "point of view" of the universe. Finally, the worth of persons is not a value to be maximized by creating as many persons as possible. It is, rather, a value that attaches to extant persons in virtue of which those persons are to be *respected*.

On this account, respecting a person involves respecting her as a being who has worth as just described. But how do we do that? McMahan suggests that we respect a person if and only if we show proper respect for both the person's good and her autonomous will concerning how her own life should go.[5] Even though a person's worth does not vary according to her level of well-being, showing proper respect for the person can involve taking her well-being into account. A terminally ill person who is suffering greatly does not thereby have any less worth than anyone else, but showing proper respect for her might involve taking her suffering into account by helping her to end her existence, as long as she has autonomously willed that we do so. Thus, on this account, an action can manifest respect for a person's worth even if it contributes to destroying the capacities that constitute the basis of this worth.[6] However, if we understand the Equal Worth Account as McMahan suggests, it does *not* yield the familiar idea (mentioned, for example, in Chapter 1) that respect for persons requires that we refrain from interfering with the choices of an autonomous person, unless those choices harm another. Suppose that the death of an individual who is autonomously trying to commit suicide would harm no one else. The Equal Worth Account leaves open the possibility that it would be consistent with respect for persons to prevent the individual from succeeding in this endeavor, if his overall well-being would best be promoted by living on.

What are the implications of the Equal Worth Account for our two test cases? Consider the different-age case first. We have one indivisible life-saving drug that can extend the life of a 20-year-old for many years or the life of a 70-year-old for a few. The good for each person is to survive, and, by assumption, each has autonomously

---

[4] McMahan, *The Ethics of Killing*, 256.   [5] ibid. 482.   [6] See ibid. 482–3 and 478.

chosen to try to do so. Therefore what would be respectful toward each of them is to give each an equal chance of getting the drug. In doing that we would be giving equal weight to each person's worth, regardless of the greater benefits that might be realized by allotting it straightaway to the younger person, and regardless of the fact that the older person has already had a full human life while the younger person has not.

One might object that we should rely on a randomizing procedure which, instead of giving both an equal chance of getting the drug, gives the 20-year-old a greater chance. For does she not stand to benefit a lot more from the drug than the older person? But this objection is misguided. According to the Equal Worth Account, respecting a person in the context of our cases involves both respecting her autonomous will and promoting her good (given that this is what she autonomously wants). But the requirement to promote the person's good does not vary with the amount of the good: we have this requirement regardless of how extensive that good may be. So the fact that the younger person stands to benefit more from the drug than the older person provides no grounds for giving her a greater chance to get it. It would be not be respectful to give the two persons different chances.

In our second example, we have to decide whether to save one person or five persons. We called this the different-number case. By assumption, all of the persons autonomously want to be saved, and, since they all have equal worth, we might rely on a randomizing procedure that gives equal chances to all—for instance, by tossing a fair coin to decide whether to save the one or the five. Hence all persons would have a 50 per cent chance of being saved. On this solution, we respect each person: through giving each an equal chance for survival, we give equal weight to each person's worth. If we instead straightaway save the five, one might argue, we would not show proper respect to the one not saved: for we would ignore what she autonomously wants and what is good for her by not giving her a chance to realize this good even though it is in our power to do so.

The equal chances solution implies that respect for persons would require everyone to have a 50 per cent chance of being saved even if we have only one person in one of the groups and a million in the other. Many people find this implication counter-intuitive. Of course, proponents of the Equal Worth Account might argue that in this case the greater benefit of saving a million people would justify not showing proper respect to the one person. But this reply is not entirely satisfying. It accepts that which many find implausible in benefit-maximizing views, namely that saving the greater number is justified even though it is unfair to the person whom we let die. Perhaps an alternative solution, namely one that implies that at least in some cases we can give higher chances to the greater number without disrespecting those who are not saved, would accord better with our intuitions. Thus, it might be suggested that the Equal Worth Account is compatible with giving proportional chances through, for example, a weighted lottery. On this procedure, both groups of persons might be assigned chances which are proportional to the number of members they have. The group consisting of five persons would have a five-in-six chance to be selected; the single person would have a one-in-six chance. The way we might think about this procedure is that each person has an equal (one-in-six) chance to be selected, but the chances of the five persons in the larger group are pooled.[7]

This procedure also seems compatible with the Equal Worth Account. If we give an equal one-in-six chance to each person, we show proper respect for her equal worth by giving equal consideration to her autonomous pursuit of her good. Moreover, by allowing the chances of the members of different groups to be pooled, we also show proper respect for their autonomous will since, by assumption, they would want their chances of survival to be as great as possible.

Furthermore, one could perhaps even argue that a policy of saving the greater number without any randomizing procedure is also compatible with the Equal Worth Account. Here is one suggestion how. You consider the autonomous will and good of the single person on one side and the autonomous will and good of a person on the other side. The claims which are based on their worth are equally pressing and they are equally balanced. That leaves the claims of the remaining four. Since these claims remain, they determine that we should save the greater number—in our example, the group of five. If the additional claims of these four people could not tip the balance, they would be ignored, and we would fail to respect these persons.[8] The single person who is not saved cannot complain that her autonomous will and good are considered less than those of others: her worth is taken into account just as much as anybody else's, even though it is neutralized by its being balanced against that of another, one might assert.[9]

Consequently, the Equal Worth Account seems to be compatible with at least three conflicting procedures for determining whom to save. Perhaps we can choose between them on the basis of further moral argument. But since the Equal Worth Account has counterintuitive implications in different-age cases, we have reason to question its plausibility.

---

[7] See Frances Kamm, *Morality, Mortality, Volume 1: Death and Whom to Save from It* (New York: Oxford University Press, 1993), 130–1.

[8] Frances Kamm calls this the Balancing Argument. See Kamm, *Morality, Mortality, Volume 1*, 116–21; see also Thomas Scanlon, *What We Owe to Each Other* (Cambridge, Mass.: Harvard University Press, 1998), 232–3.

[9] Some might nevertheless object that neither of the latter two procedures gives proper respect to the single person who has a smaller (or no) chance of being saved. Given that each person autonomously wills her chances of survival to be as great as possible, how is she shown proper respect when her chances fall short of this? Perhaps she is not. But if we accept this, then we must also accept that the single person is not shown proper respect if we give equal chances to her and to the group of five. For a 50 per cent chance of survival is not a maximal chance. But then respect for persons seems to require that everybody's autonomous wants are satisfied completely, which is impossible in the kind of cases we are considering. On this interpretation, the Equal Worth Account leaves us with either a moral dilemma or a moral gap. In other words, it entails that either we act wrongly no matter what we do, or it gives no guidance, leaving us helpless in these cases. And we could not escape the moral dilemma by not saving anyone: saving no one is surely incompatible with respect for persons.

6.2 The Respect-Expression Approach

Let us now explore the implications regarding our cases of the Respect-Expression Approach to FH. On this approach, FH amounts to the following (RFH): *Act always in a way that expresses respect for the worth of humanity, in one's own person as well as that of another.* Of course, RFH is to be understood as a categorical imperative: a principle that all of us have an overriding obligation to conform to. Kant uses "humanity" interchangeably with "rational nature," let us recall. In doing so, he suggests that having humanity involves having certain rational capacities, among which are the capacities to set and pursue ends and to conform to self-given moral imperatives purely out of respect for these imperatives. Following our practice in Chapter 2, we use the terms "humanity," "rational nature," and "capacity of rational choice" interchangeably. Rational nature is a threshold concept. If one has the set of capacities that are constitutive of it, one has it, no matter how well- or ill-developed those capacities may be.

Let us briefly remind ourselves of some of the main features of the Respect-Expression Approach. RFH commands that we act always in a way that expresses respect for the worth or, equivalently, the value of humanity. An appropriate reaction to the value of the sort that humanity has is to honor, cherish, or protect it, rather than to bring more of it about. Humanity has unconditional worth, which means that it is valuable in every possible context in which it exists and that neither what it affects nor what happens to it has any bearing on the degree of its value. Moreover, humanity has incomparable worth: it has no equivalent for which it can be legitimately exchanged. It makes no sense to say that in some context one or more instances of humanity have more or less value than one or more other instances of humanity. Everything that lacks incomparable worth has mere price, including human happiness and well-being. To say that humanity is unconditionally and incomparably valuable is to say that it has "dignity." Humanity and humanity alone has dignity, according to the Respect-Expression Approach.

In order to derive duties from RFH to act (or refrain from acting) in some way, we must rely on *intermediate premises*: premises that specify whether some sort of conduct expresses respect for the worth of humanity. An action is wrong just in case it fails to express respect for the value of humanity. But any action that fails to express respect for this value does so at least in part by suggesting an inaccurate message regarding what the value is.

As we have found in Chapter 2, it is not always easy to discern whether an action expresses respect for the value of humanity. It can be difficult to determine whether it sends a message that is compatible with the view that humanity has dignity. That is partly because an action's message can differ depending on the context in which it is performed. As I did implicitly in Chapter 2, I assume a contemporary Western context for the allocation actions tied to our cases.

Returning to our cases, suppose we give the life-saving drug to the younger patient in the different-age case on the grounds that, since she will live longer, we thereby bring about more well-being. If we do this, we express disrespect for the value of the older patient's rational nature. (This is an intermediate premise.) An action *might* express respect for the value of a person's rational nature or capacity of rational choice through expressing respect for her *exercise* of this capacity. Revisiting a previous example (2.3), suppose that a colleague has lent you money, but, as she informed you, needs the loan repaid by a certain date so she can make a down payment on a house. If you repay the loan on time, then your action expresses respect for the value of her capacity of rational choice through expressing respect for her exercise of this capacity, that is, through expressing respect for her pursuit of the end of buying a house.

But saving the younger patient would obviously not express respect for the older patient's rational nature through expressing respect for her exercise of it. For the older patient is not ready to step aside; by assumption, she is striving to survive. We do not, of course, respect this exercise of her capacity of rational choice by making a decision that in effect makes it impossible for her to attain her end.

An action might express respect for the value of a person's capacity of rational choice not by expressing respect for a particular instance of her exercising this capacity, but by expressing respect for the value of the capacity itself. Consider the action of preventing a person from committing suicide: by preserving this person's capacity of rational choice, this action would presumably express respect for it. But it would not honor her current exercise of that capacity.

In order for our action to express respect for the value of humanity, it must not suggest an inaccurate message regarding that value. Giving the drug to the younger patient on the grounds that doing so yields more well-being, however, would send a message that is incompatible with the idea that the older patient's rational nature has dignity. Just like the younger patient, the older patient has a claim to treatment. Whether or not we give it to her is not just a matter of whom we choose as an object of our beneficence. Our action would suggest that something of mere price according to the Respect-Expression Approach, namely well-being, tips the scales in favor of saving one person, the younger, rather than saving another, the older. But that message obviously conflicts with the notion that the value of humanity is incomparable.

One might suggest that we give the drug to the younger patient not on the grounds that we thereby maximize well-being, but on the grounds that the older patient has already had a full human life, that is, her "fair innings." But this action would also send a message contrary to the notion that humanity has dignity, for it would suggest that an instance of rational nature that has endured sufficiently long (making for a full human existence) has less worth than an instance that has been around for a shorter time. But this contradicts the notion that the value of humanity is unconditional, according to the Respect-Expression Approach's notion of such value. Whether a life has been long enough to constitute a full human life has no impact on its worth, for by virtue of being unconditional, the worth of any one person is equal to that of any other. And, again, the value of rational nature is incomparable: the value of a full rational existence cannot be weighed against the value of an incomplete one. So we cannot conclude that the value of the one is less than that of the other.

If we give the drug to the younger patient on either of the aforementioned grounds, then we express disrespect for the value of the older patient's rational nature. But what resource-distributing action in this difficult scenario would express proper respect for the value of each patient? It seems that flipping a fair coin to decide to whom to give the life-saving drug would do so. For this action would send the message that each patient is valuable; each is worthy of having an equal chance to be saved. And this action would also suggest that the longer existence or the greater quantity of well-being to be had by the younger patient fails to make her more valuable than the older one. The action does not seem to run afoul of the notion that persons have incomparable and unconditional value, that is, dignity.

Turning now to the different-number case, which action, if any, would express respect for the dignity of each person's capacity of rational choice? As we have seen, an action might express respect for the value of this capacity through expressing respect for one's particular exercise of it or through expressing respect for the capacity itself. But, just as in the different-age case, it is difficult to see how we could respect the exercise of the capacity of rational choice of *all* of the persons involved, given that by our assumption they are all striving to survive. Hence we will focus here on the latter mode of expressing respect.

Saving the five simply on the grounds that we thereby preserve more value would obviously send a message that clashes with the notion that humanity has dignity. For it would imply that five persons have greater value than one.

We might, of course, give each set of persons chances proportional to their number, that is, a five-in-six chance for the group of five and a one-in-six chance for the single person. One basis for doing so would be the idea that the value of each person initially entitles her to an equal chance of being saved (one-in-six), but that chance gets pooled with those of others in her group, if there are any. But adopting this procedure would likely send the message that the value of humanity is comparable. In particular, it would encourage the notion that five persons have greater value than one person. For if the basis for initially giving each person an equal chance is the value of her humanity, it is natural to think that the basis for allowing the pooling of chances among several people is the notion that several instances of humanity have more value than a single instance.[10]

The Kantian account also seems to rule out saving the greater number through "balancing" claims, at least if the basis for balancing them is the value of the persons who have those claims. Suppose we balance the claim of one member of the group of five against the claim of the single person. Since there remain four persons in the group of five whose claims have not been balanced by the claims of anyone, we save all the members in

---

[10] Jens Timmermann's "individualist lottery," which he claims is based on a broadly Kantian ethical theory, also gives proportional chances to each person, but the basis for doing so is different. See Jens Timmermann, "The Individualist Lottery: How People Count, But Not their Numbers," *Analysis* 64 (2004).

this group. In short, the claims of the five outweigh the claim of the one. But why?[11] One very natural response is to say that the five persons who are making the claims together have a value that is greater than that of the one. But if this is our reason, then by saving the five we express the idea that the value of persons is comparable.

Of course, someone might refuse to choose between saving the one or the five on the grounds that there is no way to make the choice that would avoid sending the message that persons fail to have incomparable value. She might think that adopting *any* procedure to determine which set of persons to select would suggest that some person (or group of persons) has greater or lesser value than another. But would not walking away also suggest a message inconsistent with the idea that persons have dignity? Would it not imply that no one was worth saving?

If one tries in the different-number case to proceed on the basis of the conviction that humanity has dignity (as specified in the Respect-Expression Approach), it seems that one would give each set of persons, the one and the five, an equal chance of being saved. Unlike the proportional chances and balancing procedures as we described them, doing this would not leave the impression that five persons together have a greater value than one. And, unlike refusing to save anyone, it would not suggest that no one was worth saving.

In sum, the Respect-Expression Approach is challenging to apply to our cases. It is not easy to determine what message a particular allocation procedure conveys. But in the different-age case the account has implications that clash with common considered moral judgments. For many of us believe that we ought to save the younger person, rather than to give equal chances to both. And I do not see how, on this approach, saving the younger person could express respect for the dignity of both. In the different-number case, we have not here eliminated the possibility that doing something other than giving equal chances of being saved to the five and to the one would express respect for the dignity of everyone concerned. However, this procedure seems most likely to do so. And, while it may seem unobjectionable to some, others may find it unattractive— especially given that, by the same rationale, if the set of persons we could save on one side were much larger (say, one million), then we would still have to flip a coin.

6.3 KID

In Chapter 5, I sketched a Kant-Inspired Account of the Dignity of Persons (KID):

> The dignity of persons is a special status that they possess by virtue of having the capacities constitutive of personhood. This status has several features, including the following:

---

[11] David Wasserman and Alan Strudler have pressed this sort of question in "Can a Nonconsequentialist Count Lives?" *Philosophy and Public Affairs* 31 (2003): 89–93. Like Wasserman and Strudler, I fail to find self-evident the assertion that the claim of the one person can "balance" only the claim of one among the five. Some defense of this assertion is needed. And to my knowledge no defense which is both plausible and avoids sending the message that the value of persons is comparable has been offered.

1. It is such that a person ought not to use another merely as a means. This first aspect of persons' special status is lexically prior to all of those that follow.
2. All persons have a status such that if an agent treats them in some way, then she ought to treat them as having unconditional, transcendent value.
3. The status of persons is such that if an agent treats another in some way, she ought to treat him as having an unconditional, transcendent value that does not change as a result of what the other does or of the agent's relation to him, apart from the following exceptions . . .
4. The status of persons is such that an agent ought to treat them as having a value to be respected, rather than as having a value to be maximized by bringing as many persons as possible into existence.

An agent's treatment of a person respects the dignity of that person only if it accords with the special status just described.

I omit the exceptions specified under 3 since they do not apply to our cases. The exceptions concern special relations between agents and scenarios in which someone we treat in some way is treating another merely as a means or merely as an obstacle. But the patients do not differ in the personal, social, or familial relations they bear to us. And, let us make explicit, none of the *patients* is treating another merely as a means or merely as an obstacle in a way that we can stop only if we treat that patient as having less value than the others.

No matter whom we save in our cases, we do not treat anyone merely as a means. Suppose we give equal chances of being saved to the 20-year-old and the 70-year-old, and, as chance has it, end up saving the latter. We have not treated the 20-year-old merely as a means; for we have not used her at all. We did not intend her presence or participation to contribute to our end, which is presumably something like that of distributing scarce, life-saving resources in a morally acceptable way. In our view, we could just as well accomplish that goal without the 20-year-old being around—indeed, doing so would then be far less emotionally trying. A similar point applies to the different-numbers case. If we save the five, for example, we do not use the one at all, let alone merely as a means. We do not see her presence or participation as furthering our goal of distributing life-saving aid ethically. So we need not consider the mere means constraint any further.

Although no matter whom we save in our cases we do not treat anyone merely as a means, we do count as *treating in some way* those requesting our aid. An agent treats another in some way if she intentionally does something to him in order to realize some end of hers (5.4). In giving someone a life-saving resource, we would obviously be treating him in some way. But we would also be treating in some way someone whom we chose not to save. We would intentionally be denying his request for the resource in order to promote our end, say, of distributing such resources fairly.

So the main question we face is this: which allocation decisions would treat persons as having unconditional, transcendent value? According to the concept of unconditional value invoked in KID, something has such value if and only if it has positive value in every possible context in which it exists. In contrast to the concept of unconditional value included in the Respect-Expression Approach, that contained in KID is consistent with the idea that a person's value might diminish, though not disappear (5.4). According to KID, that a particular being has unconditional worth fails to entail that the worth it possesses is of the same magnitude as that possessed by other beings with such worth. We treat persons as having unconditional, transcendent value, let us recall, if and only if, in the given context, the action we perform is among those that someone might perform if he reasonably believed his action to be (successfully and absolutely) constrained by his holding them to have this value. I will argue that saving the 20-year-old in the different age case and the five persons in the different-number case are among those actions.

As a preliminary step to seeing this, recall the example of a certain painting we hold to have exceptional aesthetic value. One way of respecting this value is to try to keep the painting in good shape by, say, protecting it from exposure to direct sunlight, and so forth. Or suppose that we hold a stand of 1,000-year-old Sequoia trees to be of special worth. One way of respecting this worth would be to do what is in our power to prevent the forest from being destroyed to make way for a mall. Acting with respect for the special value of a thing can and often does involve trying to *preserve* that thing.

Now consider the context of the different-age and different-number cases. It is our job to allocate scarce, life-saving resources among persons who, according to KID, we must treat as having unconditional, transcendent value by virtue of possessing capacities constitutive of personhood. Each of these persons has, we assume, used these capacities to set herself the aim of preserving her own life. Among the actions someone might perform if he reasonably believed his action to be (successfully and absolutely) constrained by his holding persons to have unconditional, transcendent value would be that of preserving them as best he can.

KID does not commit us to the view that respect for the value of a person *always* requires preserving her as best we can. First, KID does not rule out the possibility that failing to preserve a person might be respectful of her special status. We can, for example, imagine a situation in which a terminally ill person autonomously wills to end her life and is so bent on doing so that her rational agency is, as it were, permanently exhausted by this pursuit.[12] Practically speaking, she can pursue no other ends. Among the actions someone might perform if he reasonably believed his action to be constrained by his holding the person to have unconditional, transcendent value would be that of supporting her in her effort to die. For in this case honoring the special value of her *capacity* of rational choice might amount to respecting her one mode of *exercising* this capacity.

---

[12] With the phrase "autonomously wills" here I mean to invoke a split-level notion of autonomy of the sort that McMahan suggests. But an everyday notion would suit just as well. For distinctions between various notions of autonomy, see 1.3.

Second, an attempt to preserve a person might involve violating the constraint against treating people merely as means. Consider a terminally ill patient whose remaining life would be filled with great suffering and who, as we are aware, aims to end her life now. As long as this person has the necessary capacities, she must be treated as having unconditional, transcendent value. We might believe that doing this would involve attempting to prolong her existence as a person. For we can imagine that, unlike the person described above, this person's rational agency is not exhausted by her will to end her life: she can and does pursue other projects. Given that it is reasonable for us to believe that the patient cannot share our end of prolonging her life, our trying to do so through deceiving her, for example, would amount to our treating her merely as a means.

In the different-age and different-number cases, I have suggested that it is consistent with KID to preserve persons as best we can. But what does doing that amount to? Persons can be preserved along two dimensions. First, we can preserve a person by extending the period of time in which she possesses the capacities constitutive of personhood. For the sake of simplicity, let us say that preserving persons along this dimension involves preserving "person years."[13] Second, we can preserve persons by keeping them in existence. Again for the sake of simplicity, let us say that this involves preservation along the "person-numbers" dimension. To illustrate: if we save five people for two years, then on the person-years dimension we preserve ten years, while on the person-numbers dimension we preserve five people. Reflective common sense values the preservation of persons along these two dimensions. In the resource alloca-tion situations we are concerned with, we honor the dignity of persons if we maximally preserve them along both dimensions. Of course, there can be situations in which maximizing preservation along one dimension does not maximize it on the other. I discuss these cases below.

The implications of this account for our two test cases are relatively straightforward. Maximum preservation of personhood along the person-years and the person-numbers dimensions is consistent with KID. Consider the different-age case. Here there is no discrepancy with respect to the number of persons one can save, of course. Among the things someone might do if he reasonably believed his action to be (successfully and absolutely) constrained by his holding persons to have unconditional, transcendent value would be to give the drug to the 20-year-old. He would thereby preserve personhood for a longer period.

Consider the different-number case now. We assumed that all the persons involved are alike in all relevant respects, including their expectation of life should they be saved. It would be consistent with KID to save the five *both* because we would thereby best preserve personhood along the person-numbers dimension, and because by saving the five we would best preserve personhood along the person-years dimension. Of course,

---

[13] Extending life does not necessarily amount to preserving a person, or extending person years. An individual can be rescued from death, for example, only to live in a persistent vegetative state.

we might be faced with a different case in which the expectation of life of the five adds up to equal the expectation of life of the one person; in that case, there would be no difference along the person-years dimension. Other things being equal, we should still save the five because it best preserves personhood along the person-numbers dimension.

It is consistent with KID to save the greater number without giving equal chances for the one person and the group of five, or even without giving them proportional chances. However, the explanation for the permissibility of this action does not appeal to the idea of balancing claims; rather, it appeals to the idea of preserving personhood as best we can.

Let us now consider objections to the idea that it is consistent with KID to maximally preserve persons in the allocation cases we have in mind. First, one might object that in saving the 20-year-old over the 70-year-old we would actually be violating KID. According to KID, if a particular being has unconditional value, then its value does not increase or decrease based on its level of its health as an independent factor (5.4). But if we save the 20-year-old, we must be treating the 70-year-old as having a value that decreases on the basis of her health. For her health (or lack thereof) is presumably what will result in her earlier death, the objection concludes. In response, we need not be treating the 70-year-old as having a value that diminishes on the basis of her health. It is not her health that determines our decision, we might plausibly maintain, but rather the person-years available to her. In order to see this, suppose that the 70-year-old is very healthy, far healthier in fact than the 20-year-old who suffers from a debilitating chronic disease. We (somehow) know that the 70-year-old would maintain her high level of health until the very end, dying not of natural causes in a few years but in a sudden accident, while the 20-year-old would live for decades with her personhood intact. It would nevertheless be consistent with KID for us to save the 20-year-old.

A second concern with the idea that it is consistent with KID to maximally preserve persons in our cases comes into focus if we suppose that we must choose between saving one person for a year or another person for fifty-one weeks. Given what we have said thus far, it would seem to be compatible with honoring the dignity of persons to choose to save the first person on the grounds that we thereby best preserve personhood. But Frances Kamm argues that there are "irrelevant utilities," or benefits and harms that are not sufficient to justify a choice between life and death, since the difference between them is too small.[14] If she is correct about this, then perhaps someone who was acting under the constraint of valuing persons unconditionally would invoke the person-years dimension as a basis for choosing whom to save only if the difference between the time that can be preserved by saving one rather than another passes a certain threshold. But it is not obvious how to define that threshold.

---

[14] See Kamm, *Morality, Mortality, Volume 1: Death and Whom to Save from It*, 101–2.

One might also object that the appeal to maximally preserving persons is too vague. In particular, as presented thus far, it fails to answer the question of how two of its components, namely preservation along the person-years and the person-numbers dimensions, should be weighted relative to one another in cases in which preserving one set of persons would yield a higher value on one dimension and preserving another set would yield a higher value on the other. For instance, suppose we can save one person for eleven years or five persons for two years each. Choosing to save the one person would best preserve personhood along the person-years dimension; choosing to save the five people would best preserve personhood along the person-numbers dimension. How should we proceed?

Any proposal is likely to become complicated and controversial as we consider various combinations of different-number and different-age cases where one set of persons has a higher value on one dimension while the other set has a higher value on the other. But here is a suggestion that seems well worth pursuing. We begin by determining the proportion between the values possessed by the two sets of persons on each dimension. The set that contributes the higher value to the proportion on a given dimension is "favored" on that dimension. We then determine which proportion along the person-years and person-numbers dimensions is greater. We preserve the set of persons that is favored in the proportion that yields that higher number.

Returning to the example above illustrates this "comparative proportion procedure."[15] We have to choose between saving one person for eleven years or five persons for two years each. The one person has a higher value on the person-years dimension, but the group of five has a higher value on the person-numbers dimension. On the person-years dimension, the proportion between the values is 11/10. Thus, the one person is favored. In contrast, the proportion on the person-numbers dimension is 5/1. On this dimension, the group is favored. The second proportion is equivalent to a number (5) which is greater than that yielded by the first proportion (1.1). So, according to this method, we should preserve the group of five persons.

This procedure has intuitively plausible implications in a variety of cases, but it does generate controversial results in others. For example, suppose we can save one person for three years or two persons for half a year each. The comparative proportion procedure prescribes the first option, but many people may prefer the second option.

What seems undeniable is that people value preservation along both dimensions, although perhaps they give somewhat more weight to the person-numbers dimension. A complete weighing scheme should take this into account. Health economists have begun to carry out empirical research on the relative weights that people place on life

[15] Thanks to Paul Menzel ("Complete Lives, Short Lives, and the Challenge of Legitimacy," *American Journal of Bioethics* 10 (2010)) for suggesting this name. The procedure is described in more detail in Samuel Kerstein and Greg Bognar, "Complete Lives in the Balance," *American Journal of Bioethics* 10 (2010). A complication set aside here is that a fully developed procedure would have to be sensitive to the uncertainty that characterizes many choices with regard to the number of persons preserved and the duration of their preservation.

saving and life extension.[16] Although many questions remain unclear, such studies can go some way in helping us to solve these difficult problems. Future studies may address people's moral judgments on respecting persons in the sorts of cases that we discussed.

In the allocation cases we are considering, KID has an advantage over the Equal Worth and Respect Expression Accounts of respect for persons. KID is free of the implication that we fail to respect persons (or, more precisely, the dignity of persons) if we save the 20-year-old straightaway in the different age case. And unlike the Respect Expression Approach, KID does not imply that we fail to respect persons if we save the five straightaway in the different numbers case.

However, KID does *not* entail that we would be acting *pro tanto* wrongly if we did not straightaway save the 20-year-old or the five. As we found in Chapter 5, someone who holds persons to have unconditional, transcendent value might also hold that persons have incomparable value. Someone whose actions are constrained by these beliefs would not straightaway save the five rather than the one on the grounds that he would thereby maximally preserve the worth of persons. He would reject the idea that the worth of one person can be weighed against that of others. So he would presumably favor giving the one and the five equal chances to be saved. Of course, if one has the background belief that persons have incomparable value, then it is also consistent with KID to flip a coin in order to decide whether to save the 20-year-old or the 70-year-old.

In my view, we have reason to reject the view that persons have incomparable worth. Returning to a case we discussed in Chapter 5, suppose that a government is considering whether to build a new tunnel that would cut by three-quarters travel time between two major cities. The construction would involve blasting through granite at high altitude, and experts have instructed the government that, in all probability, in the course of the project a few workers would lose their lives. But, according to the experts, if the government goes ahead with it, in the next decade scores of traffic deaths will be avoided. Many of us believe that one reason that favors going ahead with the project is that it would be better if a few innocent people lose their lives than if scores of them do. But appeal to this reason is ruled out by the view that the value of persons is incomparable. Moreover, it is the notion that persons have incomparable value that yields the implausible result, mentioned in Chapter 2, that a soldier who jumps on a grenade to save his comrades is acting wrongly. While KID does not rule out appeal to the notion that persons have incomparable worth, it seems that many of us might want to reject this notion.

To prevent possible misunderstanding, it is important to make another point regarding KID. It does not require us always to use the comparative proportion procedure of deciding whom to save, even in cases like the different-age case. Suppose, for example, that we have to choose between saving an 80-year-old who would thrive for an additional fifteen years and saving a 40-year-old who would thrive only for an additional

---

[16] Erik Nord, *Cost-Value Analysis in Health Care: Making Sense Out of QALYs* (Cambridge: Cambridge University Press, 1999).

ten years. The comparative proportion procedure would have us save the 80-year-old; and doing that would be consistent with KID. But it would also be consistent with KID to save the 40-year-old on the grounds that since the 80-year-old has already had his "fair innings," we can legitimately treat him as having less value than the 40-year-old. It might *seem* that saving the 40-year-old over the 80-year-old would fail to treat the latter as having unconditional value. For would we not be treating him as having less value than the 40-year-old on the basis of his enjoying health that enabled him to live to a ripe old age? I believe that the answer is no. Saving the 40-year-old is among the actions that someone might perform if he reasonably believed his action to be (successfully and absolutely) constrained by his holding persons to have unconditional value. According to KID, if a being has unconditional value, then its level of health, personal satisfaction, or well-being does not *itself* (i.e., as an independent factor) determine this value. We might treat the 80-year-old as having less worth than the 40-year-old on the basis of the 80-year-old's already having had his fair share of person-years. Granted, a necessary condition for his having had his fair share is that he has had health good enough to maintain his personhood intact for many years. But that a certain level of health was necessary to get him to old age would not entail that his health itself constituted a ground for our disvaluing his continued existence relative to that of the 40-year-old. In order to see this, consider a case outside of the realm of the allocation of scarce medical resources. Suppose that someone is treating you as a mere obstacle. He is making a credible effort to hit you with a big pipe in order to eliminate you as a witness to his theft of an artwork. A necessary condition for his striking at you with the pipe is his having a certain level of health; without it, he would not be able to lift his weapon. If you kill your attacker in order to save yourself, then you treat him as having less worth than you do. But it would not be credible to say that his health itself constitutes your reason (or even one of your reasons) for treating him in this way. That a certain level of health serves as a necessary condition for a person's having a property or performing an action that warrants his being treated as having less worth than another is consistent with the person's health itself not featuring as a reason for treating him as having less worth.

6.4 KID and Benefit-Maximizing

According to benefit-maximizing views, we ought to distribute scarce medical resources in a way that produces maximum health benefit for the cost. So, for example, if we have a particular, limited budget, and with it we are able to fund one of two discrete treatment programs, each of which would have the same cost, a benefit-maximizing view would hold that we ought to go with the one that will secure the most health benefit. The quality-adjusted life-year or QALY is widely used as a measure of an intervention's health benefits.[17] One QALY is defined as one year of life in full health,

---

[17] For a helpful and brief account of QALYs, see Milton Weinstein, George Torrance, and Alistair McGuire, "QALYs: The Basics," *Value in Health* 12, Supplement 1 (2009).

while the absence of life is defined as zero QALYs. Health states during a year of life are valued on a scale ranging from those less than zero, that is, states worse than death, to 1, that is, full health. For example, suppose that one year of life without sight has a value of 0.5. In that case, an intervention that extended a person's life for two years but left him blind would generate 1 QALY, whereas an intervention that extended his life for two years of perfect health would yield 2 QALYs. There are various ways to determine the value to assign to a year of life in less than perfect health. One common method is to base a determination on "time trade-off" questions, which ask people how much of their lives they would be willing to give up in order for their remaining lives not to be lived with a certain health problem. Time trade-off questions can be posed to members of the community at large, to patients with certain conditions, or to healthcare experts.

Benefit-maximizing views would sometimes have implications that run afoul of KID. From this point on, let us invoke the benefit-maximizing view that indeed employs QALYs to measure benefits. Suppose that we, say in the contemporary United States, have just enough money to fund either a treatment program needed to prolong the life (and personhood) of one person for ten years of full health, thus securing 10 QALYs, or a program needed to improve the mobility of twelve people for the same time period, moving them from a health-related quality of life score of 0.8 to 0.9 and thus securing 1.2 QALYs. Let us assume that the increase in mobility is one from having to use a wheelchair to being able to get around with the aid of a walker. The benefit-maximizing view would have us fund the mobility-improving program. But someone who reasonably viewed his actions to be constrained by his holding rational nature to have unconditional, transcendent value would presumably not fund the mobility-improving program. In doing so, one would, in effect, fail to preserve someone's rational nature. In funding the life-preserving program, one would do no such thing. Moreover, having lower mobility than they otherwise might would presumably not prevent those not treated from effectively exercising their rational nature. Most of us can devise and effectively pursue a whole variety of projects even if we find it somewhat difficult to get around.

To invoke another example, suppose that we have just enough money to fund either a treatment program needed to prolong the life (and personhood) of eight paraplegics with a health-related quality of life of 0.7 for ten years thus yielding 56 QALYs or six people in full health for ten years, thus yielding 60 QALYs.[18] The benefit-maximizing view would entail that we save the six. But doing that would presumably run afoul of KID. Someone whose actions were, in his reasonable view, constrained by the idea that persons must be treated as having unconditional, transcendent value might save the eight or give them at least a 50 per cent chance of being saved. (She would give them a 50 per cent chance if she believed that persons have not only unconditional but also incomparable value.) She would not save the six straightaway.

---

[18] We might imagine that the six have a disorder that is more expensive to treat.

KID's implications in these cases strike many of us as plausible. Many of us believe that it would indeed fail to honor the dignity of persons and thus be *pro tanto* wrong to maximize benefits by funding the mobility-enhancing program rather than the life-saving one or by directing one's limited funds straightaway to preserving the six people in full health over the eight paraplegics.

A defender of the benefit-maximizing view might respond that, despite initial appearances, KID's implications in these cases are not plausible. The defender might argue, first, that behind a veil of ignorance each of us insofar as he was fully rational would, in order to maximize his self-interest, make the choices dictated by benefit-maximizing and, second, that the choices we would make behind a veil of ignorance in order to maximize our self-interest correspond to what it would be morally permissible for resource allocators to do.[19] But both of these claims are dubious.

Regarding the first claim, suppose you know that you will be one of the fourteen patients, all of whose lives are at stake, but you do not know whether you will be among the six who, with treatment, will return to perfect health or among the eight who, with treatment, will remain paraplegic. Assuming that you have an equal chance of being any of the persons concerned, which allocation would, rationally speaking, maximize your self-interest? The response to this question would presumably depend on what is central to your well-being. If, for example, playing soccer has been crucial to your happiness, then you might be rationally committed to opting for the allocation that would save the six. But if your main interests lie elsewhere, such as in intellectual pursuits or in your relationship with your children, then you might be rationally compelled to embrace the allocation that would save the eight, for doing that would, of course, maximize your chances of surviving. Suppose next that you know you will be one of the patients in our other case, either the one whose life is at stake or among the twelve who have limited, but improvable, mobility, but you do not know whether you will be the one or among the twelve. Assuming you have an equal chance of being any of the thirteen persons involved and are rationally constrained to maximize your own interests, which allocation would you choose? Even recognizing that you have merely a 1/13 chance of being the person whose life is at stake, must you opt for the allocation that would improve slightly the mobility in the twelve? I venture that many of us would say no on the grounds that a gain in mobility would not be worth the risk of not being able to live on. It is doubtful, to say the least, that rational, self-interested decision making behind a veil of ignorance would in our cases accord with the dictates of benefit-maximizing.

But even if it would, that would not entail that such decision making reflects what we, upon reflection, take to be morally permissible action. Suppose, for example, that there are two people who with organ transplants would live a long time, but without them would die. And there is one person who, if left alone, would also live for a long

---

[19] Claims of this sort are defended by Peter Singer et al., "Double Jeopardy and the Use of QALYs in Health Care Allocation," *Journal of Medical Ethics* 21 (1995).

time, but who has organs, say a heart and lungs, suitable for transplantation into the two. The one is, however, unwilling to be a donor; for without the organs he would die. Suppose that you must choose between a scenario in which you treat the one merely as a means for organs, thereby killing him, without anyone ever finding out that you did, or a scenario in which you and the other patient in need of a transplant die. If you know that you will be one of the three persons, but not which one, it would, rationally speaking, presumably be in your interest to choose that the one be treated merely as a means. If you survive you would suffer from remorse. But, let us suppose, you would nevertheless lead a life stocked with achievements, pleasurable experiences, and significant relationships with others, that is, a life well worth living in terms of your own well-being. Many of us nevertheless believe that it would be morally wrong for you to choose that the one be treated merely as a means. That in some situation everyone with the sole aim of maximizing his own well-being would, rationally speaking, choose something does not itself entitle us to conclude that choosing it would be right.

KID plausibly condemns as failures to honor dignity some allocations demanded by the benefit-maximizing view. There are, however, what many take to be morally permissible allocations that are endorsed by the benefit-maximizing view, but that KID would designate to be violations of human dignity. Let us begin by examining some additional allocations that KID might plausibly designate to be such violations and then work our way to the problematic cases.

Financial constraints might force us to choose between palliating a non-life-threatening, minor skin disease in thousands of people and performing intricate, life-saving surgeries on a few. KID implies that we would fail to honor the dignity of persons if we did the former. For it attaches a special status to capacities constitutive of personhood. The few will lose these capacities (die) if they do not receive operations. But, by hypothesis, the skin disease suffered by the thousands neither threatens the existence of these capacities in them nor significantly hinders their exercise. Of course, the benefit-maximizing view might imply that we treat the skin disease. A tiny fraction of a QALY gained by thousands of people can make a greater sum than dozens of QALYs gained by a few. Here many of us would endorse the implications of KID.

KID does not always imply that an allocation that prioritizes life enhancement over life-saving fails to honor dignity. Consider a case in which we must decide between performing the innovative surgery necessary to preserve the lives of five 65-year-olds who would then have ten years at near full health or giving five 35-year-olds who have major depression but who have not responded to conventional treatment an effective (but very expensive) new drug for it. Let us assume that if the depression sufferers do not get the treatment, they will nevertheless be prevented from committing suicide and that if they do get it, it will relieve them of the main symptoms of major depression for the rest of their lives, which will span a normal length. These symptoms include "markedly diminished interest or pleasure in all, or almost all, activities most of the day, nearly every day" as well as "diminished ability to think or concentrate, or indecisiveness, nearly every day." The symptoms cause "significant distress or impairment in social, occupational, or other important areas of functioning."[20] In order to determine whether treating the depression-sufferers would run afoul of KID, we must ask ourselves whether someone who reasonably believed his actions to be constrained by his holding persons to have unconditional, transcendent value might do so. And I believe that the answer is yes. He might reason, as we suggested (5.5), that holding the capacities constitutive of personhood to have such value involves not only placing in high esteem their continued existence, but also their effective exercise. But, as the aforementioned symptoms indicate, sufferers of major depression seem largely unable to effectively exercise their rational capacities. Given that the treatment is effective and that the 35-year-olds will live much longer than the 65-year-olds, someone who placed the value in question on persons might indeed opt to treat the younger patients. Of course, the benefit-maximizing view might endorse that choice as well.

Cases in which the benefit-maximizing view appears to have an advantage over KID are ones that take the following shape: We face a choice between preserving a small number of persons and of significantly improving the quality of life of many, but in a way that does *not* bring the many from a condition in which they cannot effectively exercise their capacity of rational choice to a condition in which they can. For example, we can either preserve the lives (and personhood) of two people for thirty years each or prevent 200 young people who will have full lives from becoming paraplegic. The 200 would be better off if they did not become paraplegic. But if they did, they would nevertheless be able to set, as well as to rationally and effectively pursue, a whole range of ends. We are assuming that the society in which the 200 live provides accommodations that mitigate the effects of their inability to walk. Moreover, we are assuming that the 200 will adapt to their new situation. As seems often to be the case with such disabilities, the 200 will judge that their quality of life has diminished much less than other people, who do not have the disability, would predict.[21] According to KID, we would fail to respect the dignity of persons if we chose to help the 200 instead of the two. But many people would embrace the implications of the benefit-maximizing view, which would entail that we ought to aid the 200. Some people with Kantian intuitions might, of course, insist that the implications of the benefit-maximizing view are unacceptable. If untreated, the 200 will suffer, these people might reflect, but the two will die; and, in the given context, respecting the dignity of persons requires that we treat the latter. But what if it were not 200 people who, if left untreated, would be become paraplegic, but rather 2000 or 20,000. For most of us, I venture, there is a point at which we would see giving the greater number a large benefit as justified, all things considered.

---

[20] American Psychiatric Association, *Diagnostic and Statistical Manual of Mental Disorders DSM-IV-TR Fourth Edition* (Washington, D.C.: American Psychiatric Association, 2000).
[21] See, for example, Peter Ubel, George Loewenstein, and Christopher Jepson, "Whose Quality of Life? A Commentary Exploring Discrepancies Between Health State Evaluations of Patients and the General Public," *Quality of Life Research* 12 (2003): 599–600.

There being such a point does not render KID implausible. KID specifies ways of acting that fail to respect the dignity of persons and that are thus *pro tanto* morally impermissible. But KID does not purport to arrive at all things considered verdicts regarding moral permissibility. It leaves open the possibility that although an action fails to honor persons' dignity, it is all things considered right. However, a defender of KID must insist that, in the sort of context we are envisaging, it would indeed fail to honor the dignity of the few to forego saving their lives so that others, in whom rational nature and its effective exercise are not at risk, can receive a large benefit.

It is important to recognize that in the cases that involve the distribution of scarce medical resources an action that fails to honor the dignity of persons might be demanded, and plausibly demanded, by other moral principles or considerations. KID tries to be substantive and plausible and yet to minimize conflicts with other moral demands. But some conflicts seem unavoidable. We should acknowledge them as a perhaps unfortunate but important feature of our moral landscape.

7

Markets in Kidneys

Developments on the ground and in theoretical debate underscore the importance of thinking more about the morality of buying and selling organs. "Transplant tourism" has become a popular means by which patients in need of organs procure them, and increasing numbers of bioethicists and philosophers are advocating regulated markets in organs.

In a typical case of transplant tourism, a patient from a developed nation such as the United States, Saudi Arabia, or Israel pays a fee to have a kidney transplant abroad. A "donor," procured for the patient by a broker, awaits him. The donor or, more accurately, seller is very poor, for example, someone living below the poverty line in the Philippines, and the broker has promised to pay him in cash for his kidney.[1] The Philippines recently banned kidney transplantations for foreigners in an effort to curb the organ trade.[2] Other popular destinations for transplant tourism, including Pakistan, have passed regulations forbidding the commercialization of organs, but so far enforcement has been lax. Figures are difficult to come by, but World Health Organization research suggests that in 2005 at least 5 per cent of organ transplants worldwide involved such tourism.[3] In Pakistan, for example, up to two-thirds of the 2,000 kidney transplants done that year were performed on foreigners.[4]

The United States Congress passed a law, the National Organ Transplant Act of 1984 (NOTA), making it illegal to buy an organ on American soil. Members of Congress offered as justification the argument that such commerce is immoral. But if it is indeed wrong to buy an organ in the US, is it not also wrong for an American citizen to buy one in the Philippines, say from a laborer who earns $1 per day? Laws forbidding conduct by US citizens abroad are already on the books. The PROTECT Act of 2003 makes it illegal for Americans abroad to have commercial sex with anyone under age 18. If the commercialization of organs is morally wrong and partly for that

---

[1] K. Bramstedt and J. Xu, "Checklist: Passport, Plane Ticket, Organ Transplant," *American Journal of Transplantation* 7 (2007).

[2] Carlos Conde, "Philippines Bans Kidney Transplants for Foreigners," *The New York Times* April 30, 2008.

[3] Shimazono, Y., "The State of the International Organ Trade: A Provisional Picture Based on Integration of Available Information," *Bulletin of the World Health Organization* 85 (2007): 959.

[4] Shimazono, "The State of the International Organ Trade," 957.

reason legally prohibited in the US, perhaps it also should be illegal for Americans to purchase organs abroad.

What drives transplant tourists is clear. More than 88,000 people are now on the waiting list to receive a kidney in the United States.[5] But there is a shortage of organs. In 2010, only 16,899 kidney transplants were performed in this country.[6] Some of those currently on the waiting list will die as a result of no organ being available to them. Even those who are relatively fit have an understandable desire to hasten transplant; the quicker they receive a kidney, the less time (if any) they will spend on dialysis and the longer and healthier they are likely to live.[7]

Some philosophers and bioethicists have reacted to the organ shortage and/or transplant tourism by advocating the establishment of a regulated market in organs.[8] Proposals for regulated markets vary, but they typically involve provisions that try to ensure the safety of sellers and recipients, as well as transparency regarding risks, payment, and follow-up care.[9] Some proposals specify that it be the government, rather than, say, a private clinic, that actually purchases the organs and that it provide organs and associated care to citizens in need, regardless of income. Of course, if buying and selling organs is morally wrong, then there is reason to reject such proposals.

Opponents of commerce in organs sometimes appeal for justification of their views to Kant's Formula of Humanity (FH) and the related concept of the dignity of persons. Kant implies that anyone who sells an integral part of his body violates FH and thereby acts wrongly.[10] Although these opponents' appeals to Kant's views are apt, they are less helpful than they might be. The opponents invoke the necessity of respecting the dignity of persons without specifying in detail what dignity is or what it means to respect it, and they cite the wrongness of an agent's treating another merely as a means without clarifying conditions under which this occurs.[11]

---

[5] "Overall by Organ: Current U.S. Waiting List," Organ Procurement and Transplantation Network (OPTN), 2011.

[6] "Transplants in the U.S. by State," Organ Procurement and Transplantation Network (OPTN), 2011.

[7] Madhav Goyal et al., "Economic and Health Consequences of Selling a Kidney in India," *Journal of the American Medical Association* 288 (2002): 1589. Like other medical procedures, transplants can cost far less abroad than in the US.

[8] See, for example: Janet Radcliffe-Richards, A. Daar, and R. Guttman, "The Case for Allowing Kidney Sales," *Lancet* 351 (1998); Benjamin Hippen, "In Defense of a Regulated Market in Kidneys from Living Vendors," *Journal of Medicine and Philosophy* 30 (2005); Stephen Wilkinson, *Bodies for Sale* (London: Routledge, 2003); James Taylor, *Stakes and Kidneys* (London: Ashgate, 2005).

[9] See, for example, Arthur Matas, "Why We Should Develop a Regulated System of Kidney Sales: A Call for Action," *Clinical Journal of the American Society of Nephrology* 1 (2006): 1129 and Taylor, *Stakes and Kidneys*, 110–12.

[10] Immanuel Kant, *The Metaphysics of Morals*, trans. Mary Gregor (Cambridge: Cambridge University Press, 1996), 423. I am referring to Preussische Akademie (vol. VI) pagination, which is included in the margins of the Gregor translation. I cite the *Metaphysics of Morals* as MS.

[11] See Cynthia Cohen, "Public Policy and the Sale of Human Organs," *Kennedy Institute of Ethics Journal* 12 (2002) and Mario Morelli, "Commerce in Organs: A Kantian Critique," *Journal of Social Philosophy* 30 (1999). Based on a reading of Kant, Mark Cherry contends that the sale of organs violates Kant's categorical imperative only if at least one of the following three criteria are fulfilled: the sale puts life in danger, these organs "are equivalent to oneself as the subject of morality in one's own person," or the sale "is not associated

This chapter applies some of the Kantian principles, inspired by FH, that we discussed in Part I to the issue of the moral permissibility of "live donor" transactions, namely ones in which, in exchange for money, someone chooses to undergo a kidney extraction. Let us call this sort of transaction "market exchange" of a kidney.[12] The chapter asks whether market exchange of kidneys involves agents treating others merely as means and thereby fails to respect their dignity (7.1). In order to answer this question, the chapter invokes two accounts of treating others merely as means developed in Part I. The chapter contends that many, but not all, market exchanges of kidneys involve an agent's treating another merely as a means and thus acting (*pro tanto*) wrongly. If, while engaging in a market exchange of an organ, someone treats another merely as a means, then he fails to respect the other's dignity, according to the account of dignity (KID) sketched above (5.4). But KID specifies circumstances in which someone does not treat another merely as a means yet nevertheless violates his dignity. The chapter examines whether such violations occur in market exchange of organs (7.2). It finds that they sometimes, but do not always, occur.

The chapter next asks (7.3) whether FH, interpreted in accordance with the Respect-Expression Approach discussed in Chapter 2 (i.e., RFH), sustains Kant's universal condemnation of market exchange. The chapter argues that it does not. It is helpful to see what the implications of RFH are regarding market exchange of organs, especially since many with Kantian leanings would, I suspect, take it as an important advantage of RFH if, contrary to what I argue below, it supported a universal condemnation of such exchange. A second reason to look back at RFH is that trying to discern its implications regarding market exchange will illustrate a heretofore unexplored way in which an action might express disrespect for the dignity of humanity, namely by tending to promote the idea that some persons lack dignity. And finally, developing a charitable reconstruction of why an advocate of RFH might condemn market exchange of organs will help us to recognize a morally problematic feature of some, though by no means all, such exchange.

The chapter examines market exchange of kidneys not only in light of the Kantian principles sketched in Part I and of (one version of) orthodox Kantianism, but also in light of a broadly Kantian notion more familiar in contemporary bioethics, namely the notion that autonomy has intrinsic value (7.4). Doesn't valuing autonomy lead us to support establishing markets in organs? Would markets not promote persons' ability to

---

with a discharge of a duty" (*Kidney for Sale by Owner* (Washington, D.C.: Georgetown University Press, 2005), 135). Cherry then argues that none of these criteria need be fulfilled in a given sale of an organ. "In principle," says Cherry (136), "Kant should not have an objection to selling organs when the risk to life is de minimis and when it is to discharge a duty, such as to care for one's family." According to the reconstruction of one version of the categorical imperative (the Formula of Humanity) that I consider (7.3), the sale of organs might violate this principle without fulfilling any of the criteria Cherry mentions.

[12] Of course, there might also be buying and selling of cadaveric kidneys as well as other sorts of monetary exchange involving kidneys. But by "market exchange" here I mean to focus solely on cases in which, for money, a living person sells his kidney for extraction while he is alive and hopes to recover from the operation.

act autonomously? The chapter argues that if one, in a Kantian spirit, values autonomy, then one would actually be wary of markets in organs.

In sum, market exchange of kidneys, even legal and regulated exchange, would often be morally wrong, I argue. It would often be wrong even though it might not only be consensual, but also save lives and reduce suffering by increasing the number of kidneys available for transplant. Yet I do not claim that market exchange is wrong in every possible context. If political, economic, and social conditions were different—if the world were closer to Kant's kingdom of ends than to Hobbes' state of nature—then market exchange might typically accord with moral requirements. In any case, since market exchange of kidneys would pose significant moral danger in the world as we know it, it makes sense to consider alternative means of reducing the current shortage of organs for transplantation. I do so very briefly at the chapter's end (7.5).

In order to facilitate our application of principles in the chapter, it will be helpful to have some examples in view.

Entrepreneur

A 30-year-old man who is in good health sells one of his kidneys for $50,000 in order to finance a new business. He has a secure job, but not enough savings to get his venture off the ground. The buyer is a 45-year-old woman in end-stage kidney failure whose health is deteriorating to the point that her doctors soon will deem her too sick to remain on the national waitlist for an organ. Both buyer and seller have a thorough understanding of the health risks each will incur. The buyer pays for the organ with her own funds, without giving up her financial well-being or that of her family. She has excellent insurance, which covers all the rest of her medical expenses. She pays the seller's medical expenses and guarantees him health insurance for the rest of his life. Both are cared for by first-rate physicians in a state-of-the-art facility.

Transplant Tourism

A 25-year-old, married man in a developing country has struggled as a laborer to make ends meet. Expenses for food, housing, and, especially, medical care for his wife have landed him in debt. His creditors are harassing him to pay what he owes. He agrees to sell one of his kidneys to a broker for $2,500. As the seller is aware, the broker is an agent of a nearby clinic, one that provides services to transplant tourists from wealthy countries. The broker tells the seller that it is easy to explain the risks he will face: the surgery necessary to harvest the kidney poses little threat to his well-being; in all likelihood, he will be fine in a few weeks. In addition to giving the laborer the cash, the broker pays for the surgery, as well as for the medical expenses incurred during the seller's recovery. The surgeon is experienced and competent, and the facilities for surgery and recovery are adequate.

Regulated Market

As in the previous case, a 25-year-old, married man in a developing country has struggled as a laborer to make ends meet. Expenses for food, housing, and, especially, medical care for his wife have landed him in debt. His creditors are harassing him to pay what he owes. But in his country, the government has established a regulated market in organs. A government employee gives the man a thorough and comprehensible description of the short- and long-term health risks posed by kidney extraction as well as of the benefits he will receive if he sells an organ, including $2,500 and health insurance for life. The man goes ahead with the procedure and indeed receives the cash and insurance coverage.

7.1 Treating Kidney Vendors Merely as Means

In this section I bring two of the accounts of treating others merely as means developed in Part I to bear on the question of whether the organ buyers or sellers in our examples treat one another in this way. If they do, then they fail to respect the dignity of persons, according to KID; for one way to fail to do so is to treat someone merely as a means (5.4).

Let us first ask whether the Hybrid Account (3.5) implies that any of the buyers or sellers treat one another merely as means. According to that account, an agent treats another merely as a means if she uses the other and it is reasonable for her to believe neither that the other can consent to her use of him nor share her proximate end in using him.

The Hybrid Account does not imply that anyone in Entrepreneur treats the other merely as a means. The two parties use one another. The cash-hungry entrepreneur treats the wealthy patient as a means for obtaining money; the patient treats the entrepreneur as a means for getting a kidney. But each party has the opportunity to avert being used by the other simply by withholding consent to this use. So each can consent to the other's use of him/her.

In Transplant Tourism, the laborer does not use the broker merely as a means. Nothing prevents the broker from refraining, if he wishes, from agreeing to buy the laborer's organ. However, in a real-world manifestation of such a case, the broker probably would be treating the seller merely as a means. Poor kidney sellers are very often harmed by the transaction. According to a study conducted in Chennai, India, for example, sellers were very poor, with more than half living below the poverty line before selling a kidney. Their principal motive was to rid themselves of debts, primarily ones incurred by paying for food, housing, dowries, and medical care. Over 85 per cent of the sellers reported that organ removal was followed by a decline in health. One half reported persistent discomfort at the incision cite and one third back pain—likely more than mere inconveniences, given that the majority of sellers work as laborers and street vendors. But the procedure did not deliver them from debt. Significantly more individuals lived below the poverty line after the sale than before it. Almost 80 per cent of those in the study would not recommend that others in similar circumstances sell a kidney.[13] A more recent survey of 239 kidney vendors in Pakistan showed that 88 per cent made no economic improvement, and 98 per cent reported deterioration in general health.[14] Vendors in Pakistan have also been found to be at high risk of

---

[13] Goyal et al., "Economic and Health Consequences of Selling a Kidney in India."
[14] Syed Naqvi et al., "A Socioeconomic Survey of Kidney Vendors in Pakistan," *Transplant International* 20 (2007).

developing renal disease in the long term.[15] Finally, kidney sellers, like prostitutes, suffer from a stigma that can extend to their families.[16]

In many contexts in which transplant tourism occurs, indeed in all of which I am aware, it would be remarkable ignorance indeed if a broker were not cognizant of possible harms such as these. Let us suppose that he is cognizant of them. It is, then, reasonable for the broker to believe that the seller lacks the opportunity to dissent to the particular use he plans to make of him. In order to have such an opportunity, the seller must have a basic understanding of what that use is. He must be aware of how the buyer intends to use him, what the likely effects on him (the buyer) will be, and what proximate purpose the agent intends to use him for. But the broker has obscured the character of his use of seller. He claims to disclose the risks of the transaction to the seller, but fails to mention ones like we just enumerated. It is not reasonable for the broker to believe that the seller can consent to his using him.

If it is also not reasonable for the broker to believe that the seller can share his end, then, according to the Hybrid Account, he is treating the seller merely as a means. And, in at least some real-world manifestations of the case, it would not be reasonable for the broker to believe this. For example, the broker would presumably be aware that kidney sellers have as an end to maintain their health. And he might be aware that those from whom he buys kidneys tend to experience declines in health. It would not be reasonable for such a broker to believe that the seller can share his end. For it would be reasonable neither for him to believe that the seller could, without willing to be thwarted in the pursuit of his ends, pursue his (the broker's) end of getting a transplantable kidney from him, nor for the broker to believe that if the seller was aware of the likely effects of his (the broker's) pursuit of his end, the seller would abandon his end of maintaining his health. Depending on circumstances on the ground, the buyer in Transplant Tourism would be treating the seller merely as a means.

What does the Hybrid Account imply regarding Regulated Market? In this example, let us assume, kidney sellers receive quality follow-up care that well-nigh ensures that the kidney extraction does not result in a decline in their health. The account does not imply that the laborer who sells his kidney to pay his debts treats the government (or its agent) merely as a means. It is open to the government to choose not to purchase a kidney from him. But does the government treat the laborer merely as a means? It is certainly using him. That fact, coupled with the fact that his circumstances preclude good alternatives to selling his kidney, might seem to entail that the government is using the laborer merely as a means.

But *these particular* facts do not entail this. One agent can use another, yet not treat him merely as a means even when, as they both know, the other's welfare will diminish

[15] Syed Naqvi et al., "Health Status and Renal Function Evaluation of Kidney Vendors: A Report from Pakistan," *American Journal of Transplantation* 8 (2008).

[16] Lawrence Cohen, "Where It Hurts: Indian Material for an Ethics of Organ Transplantation," *Daedalus* 128 (1999).

unless the other allows the agent to use him. To revisit an example from Chapter 4, suppose that someone's car has overheated in a desert. She might not have any good alternative but to pay the proprietor of the local service station to take her and her car to the nearest town. That the proprietor uses her to make money does not, according to the Hybrid Account, entail that she cannot consent to his use of her. Although she will not do so, she is able to avert his use of her by withholding her agreement to it.

But now suppose that the proprietor is partly responsible for her engine's overheating. At his service station 40 miles back, he agreed to check her radiator fluid, but chose not to because he was in a hurry. In that case, in proposing a repair for a fee, he is treating her merely as a means, intuitively speaking. But the Hybrid Account does not register this point.[17] That is one reason why we needed to develop additional accounts of treating another merely as a means.

One additional account is the Induced Vulnerability Account. It holds roughly that an agent uses another merely as a means if it is reasonable for the agent to believe that she has contributed to the other's being in the position that unless the agent herself uses him, his well-being will diminish significantly; it was foreseeable to the agent as well as avoidable that she would contribute to his being in this position; and she was able to, but did not, give the other an opportunity to dissent from taking the risk that she (the agent) would contribute to his being in this position.

In light of this account, the government (through its agent) *might* count as treating the laborer in Regulated Market merely as a means. The government is, of course, using him. Moreover, let us make the plausible assumption that the laborer's overall well-being will diminish from its present level unless he sells his kidney. Two key questions here are whether it is reasonable for the government to believe that it contributed to putting the laborer in this situation and, if so, whether its doing so was foreseeable and avoidable. It is not hard to imagine a government's economic and/or social policies being such that the answers to these questions are affirmative. In order to fulfill the third condition in the Induced Vulnerability Account, the government must have been in position to give the laborer a chance to avoid being pushed by its policies into a scenario in which his future well-being depends on the government's buying his kidney. A government is in such a position if, for example, it can offer citizens a social safety net: a package of goods and services that prevents them from having to sell their organs, say, in order to pay medical bills. It is not hard to imagine a government's having the resources to offer a social safety net to all of its citizens, but refraining from doing so. If all of the conditions in the Induced Vulnerability Account are fulfilled, then the laborer's government is treating him merely as a means. It is doing so even if it has the laborer's informed, voluntary consent to purchase his kidney and gives him adequate post-operative care. Of course, determining whether all of the conditions are indeed fulfilled in a given situation would require empirical research.

---

[17] Of course, that does not mean that the account is false. It does not purport to capture all cases of just using people.

According to Alan Wertheimer, it is not clear that, or how, an agent who uses another in a mutually beneficial and consensual interaction can in Kantian terms fail to respect the dignity of the other, assuming that the agent has made securing the other's informed, voluntary consent to the transaction a necessary condition for going through with it.[18] Let us assume that in Regulated Market the conditions Wertheimer mentions have been fulfilled. The transaction between the laborer and the government agent has been mutually beneficial. In particular, the laborer's well-being is overall greater than it would have been had he never parted with his kidney. Moreover, the government agent made his carrying out the transaction conditional on his securing the laborer's informed, voluntary consent to it. We have just shed light on how the government, through its agent, might nevertheless have failed to respect the laborer's dignity. It might have failed to respect his dignity by treating him merely as a means, according to the Induced Vulnerability Account. Perhaps kidney sellers such as the laborer in Regulated Market do sometimes benefit in the long run from selling. But their benefitting is consistent with the buyers' failing to respect their dignity, even when the buyers have made the transaction conditional on the seller's voluntary, informed consent to it.

In any case, I do not wish to claim that regulated markets in organs, wherever or whenever they were established, would necessarily violate a constraint against treating persons merely as means. I wish only to highlight the danger that in the world as we know it such markets would do so. Of course, that some market exchange of kidneys involves treating others merely as means does not itself entail that this exchange is wrong, all things considered. The mere means constraint might get overridden by other considerations. I leave it to the reader to determine whether, according to his or her considered judgment, it does so in the cases we have discussed. In my view, it does not.

7.2 Market Exchange of Kidneys and the Dignity of Persons: KID

Chapter 5 developed a Kant-Inspired Account of Dignity: KID. KID holds that an agent's treatment of a person respects his dignity only if it accords with the special status that all persons possess, including that they ought not to be treated merely as means. Since we have just found that some market exchange of organs involves one agent treating another merely as a means, some such exchange involves a failure of one agent to honor another's dignity. Assuming we believe that this market exchange is *pro tanto* wrong, KID provides a significant moral perspective—a perspective that is lacking in accounts that hold respecting the dignity of persons to reduce to obtaining their

---

[18] Alan Wertheimer, *Rethinking the Ethics of Clinical Research: Widening the Lens* (New York: Oxford University Press, 2010), 293.

voluntary, informed consent before using them and protecting their confidentiality. When laborers get treated merely as means in cases like Regulated Market, they give their voluntary, informed consent to being used for their organs. But their dignity is violated, according to KID, even if their confidentiality is respected.

Of course, KID specifies more ways of failing to respect a person's dignity than treating him merely as a means. One also fails to do so if one treats the person in some way, but does not treat him as having unconditional, transcendent worth or, apart from certain exceptions, if he treats him as having less of this worth than some as a result of what he does or of the agent's relations to him. Examples of the latter kind of failure to honor dignity might be treating someone who has committed a criminal offense as having less worth than someone with a clean record or treating someone from a minority group one dislikes as having less worth than someone from a different group. This latter kind of failure to honor the dignity of persons does not apply to our three cases, let us assume; we can set it aside here. Of course, market exchange of a kidney might involve a failure to respect the dignity of a person even if KID does not imply that it does. KID is not (and does not purport to be) an exhaustive account of conditions under which someone fails to respect the dignity of persons.

Does anyone in our examples fail to honor someone's dignity by not treating the person as having unconditional, transcendent value? In buying a kidney and thereby maximizing her chances of extending her life, the wealthy woman is obviously treating herself as having such value. Suppose we assume, as seems to be plausible, that the risk of mortality and morbidity posed to the entrepreneur by nephrectomy is small.[19] Partly as a result of his receiving better medical care, his body will likely face far less hardship than the laborer's. If the woman also reasonably assumes that the risk to the entrepreneur is small, she is treating him as having unconditional, transcendent value. The entrepreneur is, of course, treating as having such value the woman whose life his kidney will help to preserve. And since the risk to him of nephrectomy is, in his reasonable view, small, and the success of a project important to him is, he thinks, dependent on this operation, his undergoing it does not amount to a failure by him to treat himself as having such value.

Regarding Transplant Tourism, the laborer who sells his kidney is obviously not failing to honor the dignity of the broker who buys it. Moreover, it does not seem that he is failing to honor his own dignity; for it is presumably reasonable for him to believe that his family's well-being depends on his having a nephrectomy, and, given what the broker tells him, that the risk to him of having one is low. But is the broker failing to respect the dignity of persons in the way specified above? Suppose that it is reasonable for the broker to believe that while the laborer's life might get cut short on account of his selling his kidney, the transplant tourist who ends up receiving the organ will have

---

[19] Data on nephrectomies performed in the United States (and presumably under excellent conditions) tend to confirm this assumption. See, for example, Dorry Segev, "Perioperative Mortality and Long-Term Survival Following Live Kidney Donation," *The Journal of the American Medical Association* 303 (2010).

his life preserved by at least as many years as the laborer will lose. In that case, it does not seem as if the broker is failing to treat persons as having unconditional, transcendent value. Someone who in this context reasonably believed his actions to be (successfully and absolutely) constrained by his holding them to have this value may purchase the laborer's kidney. As we noted (5.5, 6.3), someone who reasonably believed his actions to be constrained in this way might try in his treatment of others to preserve person-hood to the best of his ability. And as far as the preservation of persons is concerned, the organ transaction we described would presumably have no negative effect.

But now suppose something different: it is reasonable for the broker to believe that, although the laborer is only 25, nephrectomy will result in his dying prematurely: after only a few more years, instead of a few more decades. And the patient in whom the laborer's kidney will be transplanted, who is elderly, will have his life preserved by only a few more years. In that case, it is questionable whether the laborer would be treating persons as having unconditional, transcendent worth. Why would someone who in this context reasonably believed his actions to be (successfully and absolutely) con-strained by his holding them to have this value engage in a transaction that would significantly diminish the net time persons would endure? Of course, in Chapter 6 we investigated in detail the idea that aiming to accord one's action with the view that persons have unconditional, transcendent value might lead one, not only to try to preserve as many persons as possible, but to try to preserve them for as long as possible.

In light of our present purposes, the case of Regulated Market does not differ enough from that of Transplant Tourism to warrant detailed discussion. If it is reasonable for the laborer who sells his kidney as well as the government agent who buys it to believe that the laborer will not thereby be shortening his life or otherwise interfering with the exercise of his rational capacities, then neither one is dishonoring the laborer's dignity by failing to treat him as having unconditional, transcendent value.

7.3 Market Exchange of Kidneys and Orthodox Kantianism

FH commands that we treat persons as ends in themselves, never merely as means. As we discussed at length in Chapter 2, one prominent interpretation of this principle, namely what I call the Respect-Expression Approach, holds that it amounts to RFH: *Act always in a way that expresses respect for the worth of humanity, in one's own person as well as that of another.* Does FH, so interpreted, generate the condemnation without excep-tion of organ selling that Kant himself seems to endorse? Application of RFH to our examples reveals that it does not. Nevertheless, this application does bring to light a morally problematic feature of some organ transactions.

RFH, let us recall, purports to be a categorical imperative and the supreme principle of morality. According to the Respect-Expression Approach, the worth of humanity, that is, its dignity, is unconditional and incomparable. It is a value not to be maximized, but rather to be respected. The unconditional worth of a person is beyond price: it is neither greater nor less than that of any other person, and it does not decrease no matter what she does or what happens to her. The incomparable worth of a person is a value that it is never legitimate to exchange, even for the worth inherent in several other persons. An agent's action respects the dignity of persons if and only if it expresses proper respect for this worth. In order to do that, the action must not express a message that is incompatible with the notion that persons have dignity. We encapsulate our judgments as to whether an action expresses proper respect for the worth of humanity in "intermediate premises." An example of such a premise that Kant would presumably endorse is the following: "Killing oneself in order to avoid pain expresses disrespect for the worth of one's humanity."

According to Kant, it is a duty to oneself not to sell one's integral parts. Might an application of RFH yield this conclusion? Kant writes:

> To deprive oneself of an integral part or organ (to maim oneself)—for example, to give away or sell a tooth to be transplanted into another's mouth, or to have oneself castrated in order to get an easier livelihood as a singer, and so forth—are ways of partially murdering oneself. But to have a dead or diseased organ amputated when it endangers one's life, or to have something cut off that is a part but not an organ of the body, for example, one's hair, cannot be counted as a crime against one's own person—although cutting one's hair in order to sell it is not altogether free from blame.[20]

Let us focus on Kant's rather odd-sounding claim that it would violate a duty to oneself to sell a tooth to be transplanted into another's mouth.[21]

In late eighteenth-century Europe, rich people would sometimes purchase live teeth from the poor. For very high fees, surgeons would extract the teeth and implant them into their customers' mouths, trying, apparently with some success, to get them to take root in their new environment. The customers purchased the teeth largely for aesthetic reasons; white, healthy teeth were in fashion.[22] Suppose that Kant had this practice in view and believed, furthermore, that the poor sold their teeth voluntarily (e.g., not under any threat of sanction if they refused) in order to increase their comfort.

Possible arguments, which appeal to RFH, for his condemnation of selling a tooth then come into view. According to one, selling and thus voluntarily undergoing an extraction poses a significant risk to the seller's health. His subjecting himself to this procedure suggests that it is worth putting one's life and thus the existence of one's humanity in danger for the sake of increasing one's comfort. But this suggestion clashes with the notion that humanity has incomparable value, value that can never be

---

[20] MS 423.

[21] I do not here address Kant's suggestion that it is wrong to give away an integral organ. But neither the Mere Means Principle, nor KID, nor RFH would even approach licensing an unconditional condemnation of such an action.

[22] Mark Blackwell, "'Extraneous Bodies': The Contagion of Live-Tooth Transplantation in Late-Eighteenth-Century England," *Eighteenth-Century Life* 28 (2004).

legitimately exchanged for something of mere price such as comfort. In short, the action expresses disrespect for the value of humanity. This argument might harmonize with the notion that to sell a tooth is to partially murder oneself. In selling it, one does not kill oneself, but one (supposedly) puts one's life at serious risk for a purpose that does not warrant it.

An obvious retort is that, even if the extraction of a healthy tooth posed a significant health risk in Kant's time, it would not now pose such a risk to most of us. So if the argument captures Kant's rationale for the conclusion that everyone has a duty not to sell his tooth, then this rationale is inadequate. It would not support the conclusion that readers of this book have such a duty.

A second argument Kant might unfurl in order to show that selling one's tooth violates a duty to oneself is the following. Granted, selling a tooth does not necessarily pose a significant risk to the health of the seller. But a tooth plays a role in the functioning of a person's body, and its location within the mouth renders it an intimate part of him. Now suppose that in order to increase his comfort a person sells a tooth. This action would demonstrate his willingness to make even intimate parts of his body available, for the right price, for others to use as they will. It would express the notion that he himself (i.e., his humanity) is available for the right price for others to employ in pursuit of their goals. This notion would constitute part of the "meaning" of his action. But the notion clashes with the idea that his humanity has dignity, for part of what it means to have dignity is to have value that is not legitimately exchangeable for anything with mere price. The action which conveys the notion thus fails to express respect for his humanity's dignity. (The last several claims together constitute an intermediate premise.) Therefore, the person's selling his tooth violates RFH.

In response to this argument, it is natural to question the claim that a person's selling a tooth for his own comfort would express the notion that he himself (i.e., his humanity) is available for the right price for others to use as they will. The message an action expresses can vary with historical and cultural context. In one context, someone's covering his face in dirt might express the message that he is unworthy of the love of God (in heaven), but in a different context it might suggest the notion that he is worthy of intimate contact with God (who is one with the earth). But in the context shared by readers of this book, as well as in Kant's context, selling a tooth is obviously not the same thing as, say, selling oneself into slavery. Someone's doing the latter in order to make a relative more comfortable would presumably express the notion that he is available for the right price for others to use as they will. This notion would constitute part of the "meaning" of his action. It would be false for this person to contend that nothing he had done sent the message that his humanity had mere price. However, it would be an exaggeration to hold that, even in Kant's time, part of the meaning of someone's selling his tooth for comfort is that he, that is, his person, is thus available. As Kant and his contemporaries were presumably aware, a tooth is not a person.[23] And, to my knowledge, selling one's tooth was not, by convention, an invitation to put oneself up for sale. In Kant's context as well as in ours, I think it would be plausible for someone who had just sold a tooth to increase his comfort to insist that in doing so he had not expressed the notion that he himself had mere price. Even if I am incorrect about this, there are surely contexts in which someone's selling his tooth would not send the message that he himself has mere price. So this second argument fails to ground a universal condemnation of selling one's tooth.

Something like Kant's concerns might register in a third argument we can construct on his behalf. Whereas the second argument maintained that a person's selling a tooth actually conveyed or expressed the notion that he himself (i.e., his humanity) or that of those like him was available for the right price for others to use as they will, this argument maintains that, at least in Kant's cultural context, actions of this type *tend to encourage or promote this notion*. An action can presumably express a notion even if it does not tend to encourage it. Someone's getting a swastika tattooed on his arm might express the notion that Jews are unworthy of respect. But, depending on the context, this action might not at all tend to encourage this notion. Actions of this kind might just be thought of as pathetic or indicative of mental illness. Moreover, an action can presumably tend to encourage a notion even if it does not express it. To illustrate with the help of an analogy: when a contemporary African-American rap artist refers to himself as a "nigger" in his songs, his doing so might not express any notion that he is unworthy of respect. It might rather reflect the complex idea that he belongs to a group that is oppressed but that has the self-confidence to appropriate a contemptuous label and use it to invoke camaraderie in the struggle against oppression. But the rap artist's use of the term might nevertheless tend to encourage the notion that African-Americans are unworthy of respect. One can imagine people, especially those who know little about rap music, thinking that in using the term "nigger" the artist is failing to respect himself. These people might then become more likely than they otherwise would be to endorse the view that the artist and his fans are unworthy of respect.

Returning to our reconstruction, the argument holds that selling a tooth to promote one's comfort is an action of a kind that tends to promote the notion that some persons have mere price. And this notion clashes with the idea that humanity has dignity, for part of what it means to have dignity is to have value that is not legitimately exchangeable for anything with mere price. But, the argument continues, an action's being of a type that tends to promote a notion that clashes with the idea that humanity has dignity amounts to the action's failing to express respect for humanity's dignity. A person's selling a tooth thus fails to express respect for humanity's dignity. (The last

---

[23] Michael Gill and Robert Sade, "Paying for Kidneys: The Case against Prohibition," *Kennedy Institute of Ethics Journal* 12 (2002): 26. In making this point, one is, of course, not committing oneself to the view that a person is constituted by something over and above his physical reality, for example, some kind of immaterial soul. There is no inconsistency in holding both that a tooth is not a person and that the capacities that constitute human persons ultimately have nothing but a material basis.

several claims together constitute an intermediate premise.) Therefore, the person's selling his tooth violates RFH.

The claim that in Kant's cultural context selling a tooth to increase one's comfort is an action of a type that tends to promote the idea that some person lacks value that transcends price might, of course, be true even if some particular actions of selling a tooth for comfort do not result in anyone's embracing (or moving closer to embracing) the idea that the seller lacks such value. But in order for the claim to be true, it must be the case that actions of the type "poor lower-class eighteenth-century European selling his tooth to augment his comfort" would frequently make someone more inclined than he otherwise would be to accept the notion that someone's humanity is available for others to use as they will. Whether the claim is true is an empirical question. But in Kant's context it is plausible. Those who sold their teeth were poor. The more affluent presumably already tended to see the poor as tools for the satisfaction of their desires.[24] It thus seems reasonable that the poor offering their intimate body parts for sale promoted the idea that they themselves (their humanity) constituted such tools.

In a context in which almost everyone has and behaves in accordance with an unshakable conviction that all persons have Kantian dignity (i.e., in a virtual kingdom of ends) it seems implausible to claim that someone's selling a tooth to secure greater comfort would be a type of action that tended to promote the idea that he lacks a value that transcends price. However, in Kant's context and in ours this claim strikes me as plausible. But it is an empirical claim, the truth of which is admittedly hard to assess. The sort of intermediate premise Kant would need to rely on in deriving a duty not to sell "integral organs" may be true in some cultural and historical contexts, but false in others. In some contexts, therefore, persons might not have a duty to refrain from selling their organs.[25]

Let us explore briefly how this third Kantian argument against selling teeth might be applied to the previous examples of organ transactions. For the sake of brevity, let us limit ourselves to the actions of the organ sellers. The question is whether the types of actions performed by the sellers would tend to promote the notion that some persons lack dignity. If so, then they would fail to express respect for the dignity of humanity and thus would be morally impermissible, according to our application of RFH.

If, according to this application, it was wrong for a poor person to sell a tooth to increase his comfort, then it seems that it also may well be wrong for the laborer in Transplant Tourism to sell a kidney to do so. The laborer already is unlikely to be seen,

---

[24]   Blackwell, " 'Extraneous Bodies,' " 51.

[25]   Kant may well reject this suggestion. After all, he appears to set out an unconditional prohibition on a person's selling an integral part of his body. His rejection would have force if it turned out that embracing the suggestion would compel one to abandon the conviction that RFH is a categorical imperative, that is, an unconditionally binding practical principle. But I don't think embracing it would compel one to do this. It is consistent to hold both that RFH allows of no exceptions, that is, that one must act *always* in a way that expresses respect for the worth of humanity, and that a particular type of action might express respect for the worth of humanity in one cultural/historical context, but fail to do so in another.

at least by some, as equal in worth to more prosperous individuals. It does not seem a stretch to maintain that his kidney vending would tend to encourage the notion that laborers themselves, and not merely parts of their body, can be had for a price and so lack Kantian dignity. It is easy to envisage a slide from the thought that these poor persons' intimate body parts are for sale to the idea that they themselves are fungible. Just how prevalent such a slide would be is a context-dependent, empirical question. A similar point applies to Regulated Market. It makes sense to imagine that the least well-off members of the laborer's society would be selling their kidneys.[26] Kidney vendors in Iran, one of the few places where a regulated market exists, tend to be in poverty and debt.[27] Unless the price of a kidney rose very high indeed, it seems unlikely that many privileged or even the middle-class citizens of most nations would choose to undergo kidney extractions for money. So the laborer's action may well belong to a kind the occurrence of which promotes the idea that some persons lack value beyond price.

The case of Entrepreneur is less clear. He sells his kidney in order to found a profit-making business. Would an action of this type promote the idea that persons have price? Since the entrepreneur is materially and socially privileged and has a wide variety of opportunities open to him relative to workers in developing countries, others would, I believe, not be predisposed to see him as a tool to be bought if the price is right. Actions of the sort he performs, a sort that involves the sale of an intimate body part, thus might not lead people to the idea that some person or persons lack dignity. On the other hand, some might focus on the materialism of the kind of project he pursues. The idea that he is willing to sell an intimate organ for profit might encourage some to think of his very humanity as something that is salable for profit.[28]

Application of RFH to our examples reveals that this principle fails to vindicate Kant's universal condemnation of organ sales. It seems plausible that in some contexts the entrepreneur's action would not promote the notion that persons lack dignity. Moreover, it is easy to imagine a context in which a well-off person's selling his kidney in order to generate funds for a famine relief agency would express proper respect for human dignity.

---

[26] Taylor, *Stakes and Kidneys*, 35.

[27] Javaad Zargooshi, "Quality of Life of Iranian Kidney 'Donors,'" *The Journal of Urology* 166 (2001).

[28] Some might object that the judgments yielded with the help of the intermediate premises constructed here are unfair to the agents in the cases. It might turn out that that the entrepreneur's action of selling a kidney is morally permissible, whereas the laborer's in Regulated Market is not. If the moral permissibility of their actions differs, its doing so will be a function of how they and people like them are perceived in light of their social circumstances. But it seems unfair to judge the laborer's action as wrong but the entrepreneur's similar action as right, since the one is no more responsible for his social class or the way others perceive this class than the other. In response, a defender of Kant (as interpreted here) might distinguish between judging an action to be wrong and blaming an agent for doing it. That the laborer's action is wrong would not entail that he is morally culpable for it, the defender might say. In particular, if it is not possible for the laborer to know that the action is wrong, then it is illegitimate to blame him for it. I discuss the relationship in Kantian ethics between the rightness of a person's action and her being morally blamable for it in Samuel Kerstein, *Kant's Search for the Supreme Principle of Morality* (Cambridge: Cambridge University Press, 2002), 162–5.

Not only does our application of RFH not yield the results some Kantians might hope for, but it also might yield false positives, that is, it might force us to condemn actions that many of us believe to be morally permissible. A charitable way of reconstructing Kant's condemnation of markets in body parts is, I suggested, to think of it as resting on RFH, coupled with two further claims. First, if an action of a particular type tends to encourage or promote a notion that clashes with the idea that persons have dignity, then it expresses disrespect for the value of humanity (and is thus wrong). Second, in many contexts, actions of buying and selling body parts do encourage or promote such a notion. Unfortunately, the first claim might lead to trouble.

We can imagine situations in which an action of a particular type would tend to promote a notion that clashes with the idea that persons have dignity and yet nevertheless be confident that that action is not wrong. For example, suppose that in a certain society convicted child molesters are not only incarcerated, but routinely allowed to be tortured by other prisoners. A journalist writes a piece critical of this situation, arguing that the convicted molesters need to be protected. But hatred of child molesters in this society is so great that actions of the sort he performed, actions of protest aimed at securing the rights of child molesters, tend to promote rather than combat the notion that some persons fail to have dignity. The actions promote the idea that there are beings "even lower" than child molesters, namely those who try to protect them from torture. (And, let us assume, the actions have no impact on persons' attitudes outside of the society.) It seems implausible to condemn as wrong an action of the sort the journalist performs. But that is what RFH, applied in accordance with the sort of intermediate premise I suggest, would do. This result is particularly ironic, since, intuitively speaking, the journalist's action seems to belong to a kind that unambiguously expresses respect for the molester's humanity.

In short, we can imagine situations in which an action of a particular type tends to encourage a notion that clashes with the idea that persons have dignity, but does so as a result of morally perverse views held by those influenced by the actions. In these situations, it seems incorrect to conclude that the actions in question express disrespect for the value of humanity and are, therefore, wrong. The example of the journalist's action is schematic, and I am unsure whether scenarios resembling it are realized. But, as a practical matter, they might be. And this possibility makes it implausible to hold that *all* actions of a sort that tend to encourage a notion that clashes with the idea that persons have dignity express disrespect for the value of humanity.

In any case, our attempt to develop Kantian grounds, based on RFH, for a condemnation of the selling of organs has yielded some insight into a possible moral problem with doing so. One does not have to accept RFH or the sort of intermediate premise I have constructed for its application to think that there would be something morally problematic in a practice that encouraged the view that, unlike their wealthier neighbors, poor citizens have the value of tools to be used at will. That there is something morally problematic about such a practice does not entail that it is always wrong. But it should, I believe, lead us to hesitate to endorse it, especially if, as I argue below, there exist alternatives to adopting it.

7.4 Autonomy and Markets in Kidneys

In discussing our examples, we have examined the moral permissibility, relative to Kantian principles, of someone's buying or selling an organ. We have found that, depending on circumstances, some of those who engage in market exchange of a kidney would be acting wrongly. But someone might wonder whether there are other, broadly Kantian, grounds for *supporting* market exchange. In this section, I ask whether a broadly Kantian view, which incorporates the idea that autonomy is something valuable in itself, supports the idea that we should construct institutions that permit regulated markets in organs.

In Chapter 1 we distinguished briefly between three notions of autonomy or autonomous action. First was Kant's own notion, according to which autonomy is the capacity of one's "proper" self, without being caused to do so by any preceding event, both to set out a law prescribing what he ought to do and to determine whether to conform to this law. Second was the "everyday" account of autonomous action, according to which someone acts autonomously if she acts intentionally, with under-standing, and free of certain kinds of external or internal control. On a third conception we considered, a person's action is autonomous if and only if she is acting on some preference of hers and, based on reflection on her values, she either does or, if she thought about it, would choose to have this preference, even in light of understanding how it arose in her. We found that if a being is a person in our Kantian sense, then she has the capacity to act autonomously in the latter two senses. In any case, it is the third notion of autonomy that we will invoke here.

In defense of regulated and legal market exchange of kidneys, one might be tempted to appeal to the idea that autonomy has intrinsic value. One might argue that if we hold autonomy to be valuable in itself then we ought, other things being equal, to embrace such exchange.[29] For allowing it promotes autonomy, both that of the buyers, whose very lives might depend on the purchase, and that of the sellers, who without money from such a sale would be unable to pursue with reasonable chances of success goals of central importance to them, such as that of securing a good education for their children. Allowing market exchange promotes autonomy in the sense that it provides options for individuals to effectively direct their lives in accordance with plans they reflectively endorse.

---

[29] James Taylor seems to make an argument that appeals to the notion that autonomy is intrinsically valuable. See, for example, Taylor, *Stakes and Kidneys*, 19, 189, and 200–1. But in a later publication, he denies that he there invokes the idea that autonomy has intrinsic value. See Taylor, "Autonomy and Organ Sales, Revisited," *Journal of Medicine and Philosophy* 34 (2009), 640–1.

One might favor market exchange of kidneys on the grounds that it promotes autonomy in the sense just described. But within the broadly Kantian framework we are considering, autonomy is most fundamentally not a value to be promoted, but rather a value to be respected. And respect for autonomy sometimes morally requires us to construct institutions so as to disallow certain autonomously chosen actions of selling, namely those that stem from an agent's having taken a "constraining option." A constraining option is one such that a person's choosing it is "likely to result in the overall impairment of [his] autonomy" or of the autonomy of other members of his group.[30] A choice results in the overall impairment of a person's autonomy if he is less able to effectively exercise his autonomy after the choice than he was before it. The option to sell oneself into slavery would presumably be a constraining option. Although one might autonomously choose to do so, say, in order to benefit one's family, taking this option is, of course, likely to impair one's autonomy in the future. One's ability to direct one's own life effectively is likely to be curtailed if one belongs to someone else. Selling a kidney can be a constraining option. In fact, it has been such an option for very poor people who sell their kidneys in unregulated markets. As the well-known study of black-market kidney exchanges in Chennai, India, mentioned above has illustrated, vendors experience a post-nephrectomy decline in health and income.[31] And it is reasonable to assume that such declines prevent them from effectively pursuing projects that they have reflectively endorsed, and thereby diminish their autonomy. Since, by offering people money for their kidneys, black markets encourage them to act on constraining options, respect for autonomy arguably demands that such markets be stopped. So if regulated markets also encouraged people to act on constraining options, then respect for autonomy would presumably also demand that they be stopped.

Some philosophers have expressed confidence, however, that *regulated* markets would not generate constraining options for organ sellers.[32] If regulations required that sellers be healthy enough at the outset to recover fully from nephrectomy, that they receive adequate postoperative care, and that they give their informed consent to the procedure, then becoming a seller would not typically diminish one's autonomy, they argue. In other words, becoming a seller would not typically curtail one's ability to pursue effectively projects that one has reflectively endorsed. If this reasoning is plausible, then regulated markets in organs would presumably respect autonomy, as well as promote it.

But the reasoning is flawed. In a regulated market just as in a black market, typical sellers would be poor, taking the only means available to them to get desperately needed funds.[33] Even if a regulated market largely forestalled the physical problems kidney sellers experienced—and, as I explain below, I think there is reason to doubt

---

[30] Taylor, *Stakes and Kidneys*, 65, 73.
[31] Goyal et al., "Economic and Health Consequences of Selling a Kidney in India."
[32] Taylor, *Stakes and Kidneys*.   [33] ibid. 35.

that it would—such a market would not necessarily prevent them from suffering psychologically. A study of kidney sellers in Iran, where there is a regulated market, has shown that vendors frequently experience feelings of worthlessness and shame.[34] They perceive themselves as akin to prostitutes and their scars as stigmata.[35] Common psychological effects of selling a kidney in Iran are anxiety and depression, which can be just as autonomy-diminishing as the physical effects of selling in a black market.[36] "Vending, especially the psychological complications, severely affected employment potential," says one researcher.[37]

Moreover, the stigma associated with kidney vending sometimes extends to members of the group to which the vendor belongs, at least if the group includes the vendor's family and village. The Chennai slum of Villivakkam got the nickname "Kidneyvakkam," as a result of many of its residents having sold a kidney.[38] A young man in Chennai complained that other boys taunted him, saying that his mother was a kidney seller.[39] Studies are needed to determine the effects of such stigma by association, but it is reasonable to worry that it might curtail the ability of those stigmatized to effectively pursue aims they have set themselves.

The autonomy of people in whose region kidney selling is widespread might be truncated in yet a further way. People sell their kidneys in an often futile effort to repay debts.[40] That has been true in black markets and would presumably also be true in regulated ones. But as a result of realizing that residents of a particular area are willing and able to sell their kidneys for cash, moneylenders might become increasingly aggressive in their debt collection.[41] Being the object of aggressive debt collection can reduce one's autonomy, it is reasonable to assume. If one is forced to raise money more quickly and in greater quantities than one would have been, one might find oneself with less opportunity to promote ends one holds to be of central importance, such as that of setting up one's own business. So for people living in a kidney-selling region, a regulated market might have autonomy-diminishing effects.

---

[34] Not many studies have addressed the well-being of Iranian kidney vendors. Malakoutian et al. ("Socio-economic Status of Iranian Living Unrelated Kidney Donors: A Multicenter Study," *Transplantation Proceedings* 39 (2007)) report that 91 per cent of vendors "were satisfied with donation" and 53 per cent suggested that others sell a kidney (825). But Nejatisafa et al. ("Quality of Life and Life Events of Living Unrelated Kidney Donors in Iran: A Multicenter Study," *Transplantation* 86 (2008)) find that after kidney extraction, the average quality of life among sellers in Tehran, measured in part in terms of physical and psychological well-being, is lower than the average among the population of Tehran (939). For general discussion of the regulated market in Iran, see Hippen, "Organ Sales and Moral Travails: Lessons from the Living Kidney Vendor Program in Iran," *Policy Analysis* 614 (2008).

[35] Zargooshi, "Quality of Life of Iranian Kidney 'Donors,'" 1795–1796.

[36] ibid. 1790, 1796.

[37] ibid. 1794.

[38] Cohen, "Where It Hurts: Indian Material for an Ethics of Organ Transplantation," 137.

[39] ibid. 140.

[40] Naqvi et al., "A Socioeconomic Survey of Kidney Vendors in Pakistan," 936; Goyal et al., "Economic and Health Consequences of Selling a Kidney in India," 1591.

[41] Cohen, "Where It Hurts: Indian Material for an Ethics of Organ Transplantation," 152.

Thus far we have been assuming that a regulated market would be an *effectively* regulated market. In other words, we have been taking it for granted that governments and businesses would largely abide by rules requiring informed consent, adequate post-operative care, and so forth. As I have just argued, even under this assumption it seems precipitate to conclude that such a market would avoid introducing autonomy-constraining options for vendors.

But how plausible is the assumption in the first place? If regulated markets in kidneys were widespread the organs would presumably flow from poor countries to wealthy ones. But poor countries tend to have poor, that is, cash-starved and ineffective, regulatory infrastructures. It seems naïve to assume that a regulated market in a very poor country would be an effectively regulated market. Government prohibitions against organ sales have been flouted in the Philippines, which has an active organ trade.[42] In India, laws on the books get ignored by corrupt officials. For example, although it violates regulations there to donate a kidney to a stranger, officials in certain areas routinely approve such donations, which are very often actually sales.[43] Might not corrupt officials also sign off on reports certifying that vendors have given their informed consent or that they are receiving adequate post-operative care? In a poorly regulated market, vendors might suffer from the same autonomy constraining effects they experience in the black market.

Granted, wealthy organ-importing nations might pass rules according to which, say, kidneys can be obtained only from countries who have embraced some international standard regarding the treatment of vendors. But even if a country embraces such a standard in good faith, rather than as a purely cosmetic measure, it might not have the resources to ensure that its citizens abide by it. Even if the broadly Kantian view of autonomy we have been considering grounds a commitment to an *effectively* regulated market in kidneys, it fails to support the conclusion that regulated markets should be established under real-world conditions.

7.5 Alternatives to Markets in Organs

Proponents of market exchange often leave the impression that embracing it is the only viable means to diminish organ shortages. But that impression is false. Many countries with organ shortages, including Australia, Germany, Great Britain, and the United States, have opt-in systems of cadaveric organ procurement. They require that potential donors give their explicit consent to donate before death. These countries might reduce their shortages by replacing their opt-in systems of donation with opt-out systems. In this sort of system, citizens are presumed to consent to donating their organs at death, but can opt out of donation if they choose. Spain, the country with the

---

[42] Sam Mediavilla, "President Wants to Stop Organ Trafficking," *Manila Times* February 8, 2007.
[43] Goyal et al., "Economic and Health Consequences of Selling a Kidney in India," 1591–92.

highest deceased-donor rate in the world, has an opt-out system in place. According to a recent analysis, even when "other determinants of donation rates are accounted for," opt-out countries have approximately 25–30 per cent higher donation rates than opt-in countries on average.[44] Former British Prime Minister Gordon Brown as well as the German National Ethics Council have called for their countries to adopt opt-out programs in an effort to reduce shortages.[45]

Presumed consent programs need to inform potential donors that (and how) they can opt out of donation. If a program fails to do this, there would be inadequate basis to presume that people who had not opted out agreed to donation. If we fail to make a robust public information campaign part of an opt-out program, we might treat donors merely as means. In taking their organs, we would be using the donors, at least according to the wide notion of using another we have adopted. And it might be reasonable for us to believe that some of the donors have neither had an opportunity to preclude this use nor are able to share our end in using them.

Against the background of a properly administered opt-out system there is another step we might take to lessen the shortage in kidneys for transplant. Currently the vast majority of organs are donated by patients who die under relatively controlled settings in the hospital—for example, by patients who die as a result of irreversible loss of function throughout the brain, but whose circulatory systems continue to work with the help of machines.[46] But organs for transplant can also be extracted from patients who die in uncontrolled settings, for example, ones who collapse at home from heart attacks and who are pronounced dead on the way to the hospital after emergency personnel have attempted to revive them. Within the context of an effectively administered opt-out system, we could reasonably presume that some of these patients have consented to donate their organs, namely the patients whose identities we can establish and who are not on record as having opted out of donation. In the United States, kidneys donated by patients who die in uncontrolled settings might lessen considerably the shortage of organs.[47]

---

[44] Alberto Abadie and Sebastien Gay, "The Impact of Presumed Consent Legislation on Cadaveric Organ Donation: A Cross-Country Study," *Journal of Health Economics* 25 (2006): 610. For a more skeptical view regarding the influence of opt-out policies on organ procurement, see Kieren Healy, "Do Presumed Consent Laws Raise Organ Procurement Rates?" *DePaul Law Review* 55 (2006).

[45] In a brief discussion, James Taylor questions both the morality and the effectiveness of a system of "presumed consent." He says that the main ethical objection to such a system is that "it will enable the state to take a person's property without his consent" (Taylor, *Stakes and Kidneys*, 8). But a system could be designed to give citizens a well-publicized opportunity to opt out of having their organs taken. In that case, the objection seems to lack force. See Michael Gill, "Presumed Consent, Autonomy, and Organ Donation," *Journal of Medicine and Philosophy* 29 (2004).

[46] Committee on Increasing Rates of Organ Donation, "Organ Donation: Opportunities for Action," (Washington, D.C.: The National Academies Press, 2006), 142.

[47] By one estimate, each year around 22,000 people who die in the United States in uncontrolled settings would be medically eligible to be kidney donors (Organ Donation: Opportunities for Action, 155–6).

It makes sense to give alternatives to markets in organs, such as opt out programs, serious consideration.[48] For the moral dangers of markets are significant. In real-world contexts organs are likely to be purchased from poor populations. I do not find good reason to believe that establishing markets would respect the autonomy of members of these populations. Indeed there is good reason to believe that doing so would introduce autonomy-constraining options for them. And there are also grounds to fear that organ purchasers might violate moral constraints by expressing disrespect for the sellers' dignity and, especially, by treating them merely as means.

---

[48] If, contrary to my expectation, robust opt-out systems fail to significantly reduce organ shortages, then we should explore other means of reducing them, besides the establishment of regulated markets. Would it be morally permissible, all things considered, to set up an "organ draft"? Such a draft might proceed (very roughly) as follows: Each year, those selected in a random drawing among citizens of a prescribed age would be screened for their physical and psychological suitability to donate a kidney. Remaining candidates would, as needed, then be required by law to give up their organ and be paid by the government a fixed price for it. Rich and poor would have equal chances of providing a resource to their fellow citizens. In any case, I do not believe that the failure of an opt-out system would itself override moral constraints proscribing the sorts of organ markets in which sellers are for the most part very poor.

8

Medical Research

Bioethicists frequently appeal to the Mere Means Principle as a constraint on research involving human subjects.[1] They affirm that when researchers use persons in their experiments, they act wrongly if they use them merely as means. This affirmation makes sense. The infamous Nazi experiments on concentration camp prisoners and the less well-known but equally brutal Japanese experiments of the same era on prisoners in China and Manchuria were wrong.[2] Part of what made them wrong, intuitively speaking, was that they involved the use of persons as mere instruments to the experimenters' ends. Any of the sufficient (or allegedly sufficient) conditions for treating others merely as means that we discussed in Chapter 3 would entail that the experimenters in these cases were doing so. Far less common than bioethicists' invoking the Mere Means Principle is their probing just what this principle amounts to.[3] In cases like those just mentioned—ones in which it is uncontroversial that experimenters are treating their subjects merely as means—employing a vague notion of what doing so amounts to seems unproblematic. However, when this morally questionable use of others might be, but is not obviously, taking place, it is helpful to have precise accounts of treating others merely as means in view. A similar point applies to the idea of dignity. Appealing to a vague idea of dignity does not lead to a mistaken notion that the experimenters in the World War II era cases mentioned failed to honor the dignity of their subjects. If these experiments do not constitute a violation of human dignity, what does? But in cases where it is not clear whether a violation of dignity has occurred, it is important to have an articulated notion of it in view.

This chapter concerns two scenarios in which it is not obvious whether the use of persons in medical experiments amounts to treating them merely as means or failing to respect their dignity. In the first scenario (8.1–8.3), researchers obtain voluntary,

---

[1] A recent example is Franklin Miller, "Consent to Clinical Research," in *The Ethics of Consent*, ed. Franklin Miller and Alan Wertheimer (Oxford: Oxford University Press, 2010), 379. See also Alan Wertheimer, *Rethinking the Ethics of Clinical Research: Widening the Lens* (New York: Oxford University Press, 2010), 47–8. Bioethicists typically do not distinguish between treating others merely as means being wrong *pro tanto* and wrong all things considered, but we can bracket that complication here.

[2] For accounts of these experiments, see George Annas and Michael Grodin, "The Nuremberg Code," in *The Oxford Textbook of Clinical Research Ethics*, ed. Ezekiel Emanuel et al. (Oxford: Oxford University Press, 2008).

[3] But see Rieke Van der Graaf and Johannes Van Delden, "On Using People Merely as Means in Clinical Research," *Bioethics* 25 (2011).

informed consent from subjects to take biological samples (e.g., blood) for use in a particular investigation. But after the subjects' personal information has been removed from the samples, the researchers give it to another group of investigators who, without the subjects' knowledge or consent, use it in a different study. The second scenario (8.4–8.6) is one in which pharmaceutical researchers obtain the voluntary, informed consent of citizens in a developing country to serve as research subjects in a trial of a new drug for a serious medical condition from which they (the subjects) suffer. If a trial of this drug were conducted in a developed country, all participants would receive effective treatment for their condition. But in the trial as it takes place in the developing country, not all subjects receive such treatment. With the help of the principles developed in Part I, the chapter inquires into whether the researchers in these scenarios treat their subjects merely as means or fail to honor their dignity.

8.1 Research on "Anonymized" Biological Specimens

Suppose you agree to participate in a study of heart disease. You give your informed consent to a group of investigators to have some of your blood used in research that aims to shed light on genetic bases of arterial sclerosis. After the investigators are finished with the study, they render your sample individually unidentifiable to anyone outside of their group. For example, they remove any information from the sample's label that would enable outside investigators to discover that the sample was yours. They also take information from your medical records and ensure that, although it is connected to the blood sample, it cannot be traced back to you. Without your knowledge or consent, the investigators then provide your "anonymized" sample and information to an outside group of investigators who use it for a study on the genetic bases of intelligence.

Neither team of investigators has violated current United States federal research regulations. These regulations specify that non-identifiable samples and accompanying non-identifiable information can be used by investigators without the knowledge or informed consent of the persons who contributed these materials and without approval from an Institutional Review Board.[4] (An Institutional Review Board or IRB is a committee charged by federal regulations to review, approve, and oversee human research in the United States.) The rationale for the regulations' permitting such use is apparently that it poses minimal risk to the contributors.[5] They have already been subject to the risks posed by venipuncture or biopsy, for example, so the only risk of

---

[4] Michelle Mello and Leslie Wolf, "The Havasupai Indian Tribe Case—Lessons for Research Involving Stored Biologic Samples," *New England Journal of Medicine* 363 (2010): 205. See also Jill Pulley et al., "Principles of Human Subjects Protections Applied in an Opt-Out, De-identified Biobank," *Clinical and Translational Science* 3 (2010): 42; Mark Rothstein, "Is Deidentification Sufficient to Protect Health Privacy in Research," *American Journal of Bioethics* 10 (2010): 7; and US Department of Health and Human Services, "OHRP: Guidance on Research Involving Coded Private Information or Biological Specimens" (Washington, D.C.: October 16, 2008).

[5] Mello and Wolf, "The Havasupai Indian Tribe Case," 205.

harm to them from the use of their materials is through their private information being released. Yet that is not a significant risk since their samples have been anonymized, or so the rationale would seem to go.

That the investigators have honored United States regulations does not, of course, entail that their actions have been morally permissible. Even if the use of contributors' materials for purposes to which they have not agreed poses minimal risk of harm to them, the question of moral permissibility remains. As I have suggested (4.4), harming another, or putting him at significant risk of being harmed, through one's use of him are not the only ways of using him wrongly. One might do neither, and yet treat the other merely as a means. Indeed, in some cases investigators' use of anonymized samples without the consent of contributors amounts to using these people merely as means, or so I will try to show. If I am correct, then there is some reason to revise United States regulations. The US Department of Health and Human Services is now considering doing just that. It has proposed reforms that would require written consent for research use of biological samples, even ones that have been "anonymized." This chapter can be read as developing a Kantian basis for such reforms (or for the new policies, if they are implemented).[6]

In Chapter 3, we developed a sufficient condition for an agent's using another merely as a means, namely the Hybrid Account:

Suppose an agent uses another. She uses him merely as a means if it is reasonable for her to believe both that he is unable to consent to her using him and that he cannot share the proximate end(s) she is pursuing in using him.

We need to ask the following: Are investigators using contributors of specimens and/or medical information? And is it reasonable for investigators to believe that contributors are unable to consent to their using them and unable to share the proximate end investigators are pursuing in using them? In cases where the answers to both questions are affirmative, investigators are treating contributors merely as means and thus acting (*pro tanto*) wrongly.

8.2 Using Biological Samples, Using Information, and Using Persons

Are investigators who use anonymized biological samples (and accompanying information) thereby using the samples' contributors? If they are, that does not itself entail

---

[6] "Reforms would require written consent for research use of biospecimens, even those that have been stripped of identifiers. Consent could be obtained using a standard, short form by which a person could provide open-ended consent for most research uses of a variety of biospecimens (such as all clinical specimens that might be collected at a particular hospital). This change would apply only to biospecimens collected after the effective date of the new rules" ("Regulatory Changes in ANPRM," US Department of Health and Human Services). See also Ezekiel Emanuel and Jerry Menikoff. "Reforming the Regulations Governing Research with Human Subjects," *New England Journal of Medicine* 365 (2011).

that they are acting (*pro tanto*) wrongly, of course. Treating another as a means is often morally unproblematic. On our account (3.1), an agent uses another if and only if she intentionally does something to or with (some aspect of) the other in order to realize her end, and she intends the presence or participation of (some aspect of) the other to contribute to the end's realization. Despite the investigators' not having access to the contributors' names, addresses, or social security numbers, they are intentionally doing something with some aspect of the contributors, namely the samples extracted from their bodies, and they intend the samples to promote their research goals. So, according to a wide construal of our account of persons using others, investigators who use anonymized biological samples taken from contributors are using the contributors.

One might raise the following worry regarding the conclusion that in using (an aspect of) a person's body, one is using the person. We seem to be committed to the view that using a person just amounts to using his rational nature. For it is a being's rational nature, including his capacity to set and rationally pursue ends, to understand the world, and so forth, that makes him a person, as opposed to some other kind of being (1.3). But using a person's kidney or a vial of his blood obviously fails to amount to using any of the capacities that make up his rational nature. So, it seems, to use a part of a person's body that is not an aspect of his rational nature is not to use the person.

However, upon reflection this concern dissipates. It does not follow from a being's having properties or capacities that distinguish it from beings of other kinds that using the being always amounts to using these particular properties or capacities. Among the properties that make something a piano, rather than, say, an organ or a table, is its having wire strings that make a sound when hit by felt-covered hammers triggered from a keyboard. Sometimes, when using a piano (e.g., by playing it), one is using this property. But one can also use a piano by using properties it possesses that are not central to making it a piano as opposed to something else. One can, for example, use it to enhance a room's décor. In an analogous way, one can use a person without using the properties or capacities the possession of which are central to making him a person, rather than some other sort of being. If I extract one of a person's kidneys so that I can sell it, I use the person, even though I do not thereby use any of the capacities that are central to making him a person as opposed, say, to some other kind of animal.

There are grey areas between treating persons as means and taking actions that are closely related to doing so. First, one might wonder whether in using *any* part of another's body, no matter how inconsequential, one treats the other as a means. One might claim that one doesn't treat the other as a means if the part one employs is inessential to the agent's functioning, or if it is external and renewable, for example. Such claims do not seem promising. A healthy person's second kidney is inessential to his functioning, but in extracting it, an organ trafficker is using him. A hair is a body part that is both renewable and external. But if a forensics instructor plucks a hair from a student's head for an upcoming demonstration of evidence-recovery techniques, then she presumably treats the student as a means. If a future scientist finds a (living) stranger's hair and uses the DNA it contains to clone him, he treats the stranger as a means, it seems. It makes sense, I think, to hold that whenever an agent intentionally does something to or with a part of another (extant) person's body and intends this body part to contribute to her realizing her end, the agent counts as using the other.

But what if investigators do not use for their research parts of an extant person's body, but rather anonymized information about the person, that is, information that the investigators cannot trace back to the person in particular? Are they using the person? In terms of ordinary English, this is a borderline case. Suppose that without his patients' consent, a physician sells to pharmaceutical companies anonymized test results contained in their charts. It is easy to imagine a patient protesting to the physician, "You're just using me," which might imply that, in the patient's view, by selling his results the physician is indeed treating him as a means. Yet it admittedly also sounds plausible to distinguish between using someone and using information *derived from* someone and to insist that, in selling the results, the physician is really doing the latter.

In general, the more detailed and unique to an individual information derived from him is, the more plausible it seems to say that in using this information one is using the person himself. To illustrate one end of the spectrum, suppose it were possible to produce a genetic replica of a person based on knowledge of the precise structure of his DNA (as opposed to producing one with actual DNA taken from the person's cells). It sounds natural to say that a scientist who in this way produced a replica of a person would be treating the person's body as a means, even though he would not actually be using any part of it. Of course, the precise structure of a person's DNA is a highly detailed and intimate characteristic, one that distinguishes him from others. To see the other end of the spectrum, suppose that, before boarding a small plane, its passengers are told by a flight attendant to step together onto an industrial scale. The passengers' total weight is then transmitted to the pilot so she can determine how much cargo it is safe to take on. It seems natural to say that the pilot is using information gleaned from her passengers. But many of us would find it odd to say that, in using this information, the pilot is thereby using the passengers themselves.

In any case, I suggest that we give a wide reading of treating another (extant person) as a means, according to which an agent counts as treating another as a means if she treats detailed information that is relatively unique to the other's body as a means.[7] Of course, information can be both detailed and relatively unique to a person's body without being identifiable by a researcher as information gleaned from the particular individual with a certain name, address, and so forth. A complete medical history spanning thirty years is detailed and relatively unique to an individual's body, but that does not entail that an investigator knows who in particular the history belongs to. As far as I can tell, there is no precise way to determine when information derived from a

---

[7] This view is not atypical. According to Franklin Miller, even when individuals' data has been de-identified, "the individuals whose data are being accessed for research are still being used for research..." Miller, "Consent to Clinical Research," 394.

person's body is individualized enough to warrant saying that in using it, we would be using the person. Judgment seems to be required here on a case-by-case basis.

Borderline cases also appear in the context of usage of rational capacities. Does one count as using a person if one uses the products of his exercise of his rational capacities? For example, might one treat a person as a means if one uses an idea he has? I suspect that the best answer to this question is similar to the one offered regarding the usage of information derived from a person's body. The more unique to a person an idea is, the more it makes sense to say that in using the idea one is using the person.[8]

Some philosophers, I acknowledge, might insist that neither in using biological samples she did not extract, nor in using personal information is an investigator using another person. Some might, in other words, reject our wide construal of using others. For example, they might hold that a surgeon who removes a person's kidney so it can be transplanted into another is using the person, but a different surgeon who does the transplantation is not using him. And some might make a sharp distinction between treating a person as a means and treating information that stems from a person, no matter how unique to him it might be, as such. I do not wish here to enter into metaphysical controversies, for example regarding when a body part constitutes an aspect of a person and when it does not. In the next section, I discuss researchers' use of biological samples such as blood. I assume that in using such a sample one is using a person, and I ask whether or when, according to the Hybrid Account, one is using him merely as a means. Philosophers who hold that in using such a sample one is not using a person would deny that the Hybrid Account even applies to the cases I examine. If, in doing an experiment on a biological sample, I am not using a person at all, I cannot be using him merely as a means. Near the end of the next section I will make some remarks that, I hope, will show that even to such philosophers our inquiries into the morality of using biological samples can be instructive.

8.3 Treating Contributors of Biological Samples Merely as Means

We are trying to determine whether investigators treat contributors of non-identifiable biological specimens or medical information merely as means. We have found it

---

[8] According to Paulus Kaufmann, an agent does *not* use another if in interacting with him she is aiming ultimately at realizing a state in the other, whether good or bad ("Instrumentalization: What Does It Mean to Use a Person?" in *Humiliation, Degradation, Dehumanization: Human Dignity Violated*, ed. Paulus Kaufmann et al. (Dordrecht: Springer, 2011)). I do not see why we should agree with this claim. Suppose, for example, that one of my ultimate aims is for a friend to get the satisfaction of winning an amateur tennis tournament. Although a very talented player, he tends to be very lazy in preparation. In order to get him motivated, I tell him (truthfully) that if he makes it to the finals he will in all likelihood be playing against a certain political consultant: one who devises inaccurate but highly effective advertisements that attack candidates my friend supports. Energized by this information, my friend practices very hard and wins the tournament. It does not sound odd to say that I used his disposition to be motivated by facing a political enemy (and thus used him) to help to realize my ultimate aim of his getting satisfaction from winning a tournament.

reasonable to assume that they are treating as means contributors of biological specimens and that they might be treating as means contributors of medical information,
depending on how unique to the individual such information is. We now need to ask
whether and/or when it is reasonable for investigators to believe that contributors are
unable to consent to their using them and unable to share the end investigators are
pursuing in using them.

If investigators do not inform contributors of their plans to use them, then it is, of
course, typically reasonable for the investigators to believe that the contributors cannot
consent to their using them. On our understanding, someone is able to consent to
being used only if he is able to prevent the use by withholding his agreement to it or to
a set of rules according to which the use is legitimate. That someone would, *if queried*,
consent to being used does not entail that he can consent to it; for he might lack any
opportunity to avert the use by withholding his consent to it or to rules that clearly
imply its legitimacy. Now a contributor cannot, practically speaking, avert someone's
use of him when no one has ever made him aware that such use will (or might) take
place. Investigators often use data contained in medical records for observational
research without patients' consent. Such research "provides valuable knowledge
regarding risk factors for disease, the safety of pharmaceuticals and medical procedures,
and the quality of medical care."[9] Depending on the detail and uniqueness of the data
they are using, investigators conducting such research are treating patients as means.
And it is obviously not reasonable for the investigators to believe that the patients can
avert their doing so by withholding their agreement to it; for the patients are (at least
typically) unaware that the research is taking place. Investigators also often do research
on de-identified biological samples without informing contributors that this research
will or might take place.

> Academic medical institutions have long performed research that involves data and/or tissue
> derived from humans, but is not human subjects' research as defined by the regulations. We
> and others have observed that some members of the lay public are unaware of the federal
> definitions and believe that existing tissues (e.g., stored surgical pathology specimens) are
> not (and should not) be used for research without explicit consent. This belief contradicts
> the current actual practice of research permitted by the Code of Federal Regulations.[10]

It is reasonable for investigators to believe that "some members of the lay public"—I
would venture many—cannot consent to the investigators using them. These people
are in no position to prevent their being used by withholding their consent to it for the
simple reason that they are unaware that the use is taking place. And they have not
agreed to rules according to which its taking place is legitimate.

---

[9] Miller, "Consent to Clinical Research," 393.
[10] Pulley et al., "Principles of Human Subjects Protections Applied in an Opt-Out, De-identified Biobank," 42.

As we found in Chapter 3, that it is reasonable for an agent who uses another to believe that the other cannot consent to this use fails to entail that she is using the other merely as a means. However, if it is reasonable for an agent who uses another to believe both that the other cannot consent to this use and that he cannot share the end the agent is pursuing in using him, then she *is* treating the other merely as a means. A key question for us is thus that of whether it is reasonable for investigators to believe that those whom they use can share their ends.

In Chapter 3, we developed a detailed account of when a person cannot share an agent's end. Part of the account was designed to prevent it from having the implausible implication that victims of coercion (e.g. mugging victims) can typically share the ends of those who coerce them (e.g., muggers aiming to get the victims' money). But the cases we are concerned with are not ones of coercion. Coercion involves threats, but investigators who use anonymized biological samples and/or detailed medical records for their research do not threaten those whom they thereby use: the investigators do not even know who contributed the samples/records. So for our purposes in this chapter it makes sense to employ a simplified account. Let us say that in the sorts of cases we are concerned with, contributors cannot share an investigator's end if:

The contributor has an end such that his pursuing it at the same time that he willed the investigator's pursuit of her (proximate) aim would violate the hypothetical imperative, and he would be unwilling to give up pursuing this end, even if he was aware of the likely effects of the investigator's pursuit of her aim.

Let us illustrate this account with the help of our original hypothetical example. You give your voluntary, informed consent to one investigator to be used as a blood donor in a study of arterial sclerosis, but you are also used by a second investigator, who, without your knowledge or consent, performs tests on your anonymized blood sample, aiming to get data for a study on genetic bases of intelligence. Suppose that you believe such studies to have potentially damaging social consequences, and you aim not to do anything to facilitate them. According to our account, you cannot share the second investigator's end. In effect, our account asks us to perform a thought-experiment. First, it asks us to imagine that you seek to realize, not only your aim of not facilitating genetic studies that have potentially damaging social consequences, but also, at the same time, the investigator's aim of gaining data from your blood for just such a study. But doing that would violate a principle of practical rationality, namely the hypothetical imperative, which commands that if you will an end, then you ought either to will necessary means to it in your power or give up willing the end. In the thought experiment, you do not give up willing your end, but, in effect, will to be thwarted in attaining it, since you also will the investigator's end. But if you will to be thwarted in attaining your end, then you fail to will means necessary and in your power to attain it. You violate the hypothetical imperative. Second, our account of end-sharing asks us to determine whether, if you knew of the likely effects of the investigator's pursuit of her end, you would be willing to give up your end. In this case, the answer is obviously no. Your gaining an understanding of the investigator's purpose would prompt you to object to what he was doing with your blood. So you cannot share the investigator's end.

That, combined with your not having been able to consent to the investigator's use of your blood, does not itself entail, according to the Hybrid Account, that she treats you merely as a means. In order for the account to entail that, it also has to be the case that it is *reasonable for her to believe* that you can neither consent to her use of you nor share her end in using you. According to the notion we have been relying on, what it is reasonable for an agent to believe is what the evidence available to the agent favors, given his education, his intelligence, his upbringing, and so forth. The investigators we are discussing are, of course, typically highly educated and intelligent; and they have access to vast libraries and refined tools for extracting information from them. Chances are, it is reasonable for the investigators to believe that at least some of the contributors of blood samples object to research on the genetic bases of intelligence and aim not to further it. The Hybrid Account does not yield the conclusion that the investigators have treated you or any other person *in particular* merely as a means. But it does imply that they are treating some unspecified number of sample contributors as such.

It might be helpful to consider another hypothetical example of investigators treating sample contributors merely as means, according to the Hybrid Account. Suppose that a group of patients have cystic fibrosis, namely an inherited disease that causes digestive problems and lung infections and that, despite recent advances in treatment, significantly curtails lifespan. The patients agree to participate in a study of treatment of the disease. They give their informed consent to a group of investigators to have some of their blood used in research that aims to study the long-term effects of cystic fibrosis patients' taking mucus-thinning enzymes to prevent lung infections. After the investigators anonymize the samples, they provide them (as well as anonymized information about the patients) to an outside researcher who, without the patients' knowledge or consent, uses them for a study on how to detect whether early-stage fetuses will develop the disease.[11]

It is reasonable, let us suppose, for the researcher to believe that the contributors of the blood were not given the opportunity to dissent from being used for her study. But it would presumably also be reasonable for her to believe that at least some of them could not share her end. For some of them would have the aim of discouraging or at least not promoting the development of very early in-utero detection of genotypes associated with cystic fibrosis on the grounds that such detection would result in an increase in the number of fetuses with such genotypes being aborted. Of course, it would be practically irrational for someone with this aim to try, at the same time, to secure it as well as the researcher's end. And such a person would obviously not give up his aim in light of an understanding of the likely effects of the researcher's pursuit of her

---

[11] Mark Rothstein offers a similar example: Rothstein, "Is Deidentification Sufficient to Protect Health Privacy in Research?" 7.

end. These effects are part and parcel of what the person aims to avoid. So the researcher would be treating some of the sample contributors merely as means.

We might find a third example of the Hybrid Account implying that investigators have treated contributors of biological materials merely as means in an actual case that has attracted attention from the American media. In 1990–1991 researchers from Arizona State University obtained blood samples from over 100 people, all members of the Havasupai Indian tribe of Northwest Arizona. Although "pre-study communications with tribal leaders apparently focused on diabetes," which tribe members develop at a high rate, contributors of samples all signed consent forms that described the research they were agreeing to participate in as studying "the causes of behavioral/medical disorders."[12] The investigators shared the samples with other investigators who used them in several studies, including ones on the genetic bases of schizophrenia, evolutionary genetics, and inbreeding. In 2004 members of the Havasupai sued Arizona State University, alleging fraud, breach of fiduciary duty, negligence, and so forth. In April 2010, Arizona State University settled the suit, agreeing to pay $700,000 to forty-one tribe members. Our concern here is not whether the investigators violated the tribe members' legal rights, but rather whether, according to the Hybrid Account, in cases such as this one investigators treat sample contributors merely as means, and thus act in a (*pro tanto*) morally impermissible way.

In order to address this concern, it will be helpful to ask a more specific question, one predicated on certain assumptions we will make about the case. The assumptions will facilitate our discussion, but for our purposes nothing important rides on whether they are true. We are inquiring into conditions under which investigators treat subjects merely as means. Our aim is not to morally condemn (or absolve) these particular investigators.

The investigators used some of the blood that the Havasupai donated for a study in evolutionary genetics, one that tries to bolster the view that a certain kind of natural selection, namely balancing selection, has been vital to maintaining genetic variation in a particular genomic region.[13] It seems implausible to claim that this study investigates the causes of behavioral/medical disorders, and on that basis to insist that the Havasupai consented to it. The study investigates how genetic variation is sustained in a genomic region that apparently plays a significant role in the immune system. But it does not focus on the relations between this region and any particular disorders. According to the study's abstract, it contributes to understanding the "evolutionary forces that have shaped human populations."[14] Such an understanding might prove useful in the investigation of the causes of illness, but an effort to contribute to this understanding does not itself constitute an investigation of the causes of illness. In any case, we will

---

[12] Mello and Wolf, "The Havasupai Indian Tribe Case," 204.

[13] Therese Markow et al., "HLA Polymorphism in the Havasupai: Evidence for Balancing Selection," *American Journal of Human Genetics* 53 (1993): 943–4.

[14] Therese Markow et al., "HLA Polymorphism in the Havasupai," 943.

assume that it was *not* reasonable for the investigators to believe that the Havasupai consented or had the opportunity to consent to the use that these investigators put them to. The investigators knew, we will suppose, that they consented solely to research into the causes of illnesses. We will also assume that the samples the investigators used were anonymized, that is, that although the investigators had access to some information regarding the sample donors (e.g., whether they were male or female), they were not aware of who, in particular, the samples stemmed from. Our specific question is whether the investigators, as described, treated some of the sample contributors merely as means, according to the Hybrid Account.

In order to answer that question we need only inquire into whether it was reasonable for the investigators to believe that all of the Havasupai they were using in their study could share the end they were pursuing in doing so. I doubt whether it is reasonable for them to believe this. On the form they signed, the Havasupai consented to participate in research on "the causes of behavioral/medical disorders." It was reasonable for the investigators to believe that at least some of the Havasupai held the following: The investigators were under an obligation, one that they apparently acknowledged by asking them to sign the consent form, to get their consent to use them for research on the causes of illnesses (e.g., diabetes). Moreover, the investigators would also be obligated to get their consent for any other sort of research they would use them for. In light of such reflection on the views of the Havasupai, reflection that seems well within the capabilities of the investigators, it seems that it was reasonable for them to believe that at least some of the Havasupai would have as an end not to have the blood they gave used for research without their consent. This conclusion is especially plausible in light of the troubled history in North America of medical research being conducted on minority populations, including Native Americans—a history of which the investigators would presumably have some awareness.[15] The research subjects could not share the investigators' end of showing that balancing selection has been vital to maintaining genetic variation in a particular genomic region, it was reasonable for the investigators to believe. For in willing their pursuit of this end, which involved use of their blood without their consent, the Havasupai in question would be willing to be thwarted in their pursuit of their end of not having the blood they contributed used for research without their consent. Moreover, it was reasonable for the investigators to believe that the Havasupai would not be willing to give up the pursuit of their end in light of the likely effects of the investigators' pursuit of theirs. As the investigators were surely aware, the Havasupai might not care at all about balancing selection, let alone enough to forgo their aim of having a say in the uses of their own blood. There is good reason to believe that, according to the Hybrid Account, the researchers treated some of the Havasupai merely as means and thus acted (*pro tanto*) wrongly.

[15] See Ron Whitener, "Research in Native American Communities in the Genetics Age," *Journal of Technology Law & Policy* 15 (2010).

Of course, according to the position defended in this book, the Mere Means Principle specifies a defeasible constraint on action. The book acknowledges that it is sometimes morally justified, all things considered, to treat others merely as means. But in the Havasupai case the constraint would surely not get overridden. The prospect of a theoretical advance in our understanding of the mechanisms of human evolution does not justify treating research subjects merely as means. In my view, an analogous point can plausibly be made in the other two cases we discussed of treating sample contributors merely as means. Even if research on the genetic bases of intelligence or on early identification of fetuses that will develop cystic fibrosis would contribute to saving a great number of lives and to reducing a great amount of suffering—prospects that seem highly questionable to me—it seems reasonable to hold that such research could be conducted, albeit perhaps at a slower pace, without treating anyone merely as a means.

In sum, the Hybrid Account yields the conclusion, which many of us find plausible, that the researchers in the cases described act (*pro tanto*) wrongly, and it points us to reasons why their action is (*pro tanto*) wrong. The researchers fail to respect the dignity of those they use in their experiments. They fail to do this by treating these people merely as means. In particular, it is not reasonable for the researchers to believe either that their subjects have an opportunity to prevent their use of them through withholding their agreement to it or that their subjects can share the ends they are pursuing in using them.

It is worth keeping in mind that the Hybrid Account does *not* condemn all non-consensual use of persons in research. For example, suppose that a trauma victim comes into a university hospital emergency room having lost a great deal of blood. The physician on duty determines that the treatment that would give him the best chance of survival is an experimental one. So she gives it to him, trying to save his life but also using him for research. It would, of course, not be reasonable for the physician to believe that her patient was able to consent to her use of him. But it would be reasonable for her to believe that he could share her end(s) in using him. Her pursuit of her ends, namely those of saving his life and producing knowledge regarding the experimental treatment, might clash with an aim of the patient's, namely that of not being used for research without his consent, the physician might reasonably believe. But it would also be reasonable for her to believe that the patient would be willing to give up (in these circumstances) pursuing this end, if he was aware of the likely effects of the physician's pursuit of her ends, namely that his chances of survival would increase. In short, it is reasonable for the physician to believe that the patient can share her end, and so the Hybrid Account would not have the implausible implication that her using him was (*pro tanto*) wrong.

In any case, it is no mystery how to prevent the use of biological samples in the sort of cases we have described from amounting to treating others merely as means, according to the Hybrid Account. Preventing this would involve giving contributors an opportunity, by withholding their consent, to avert the use of their samples in research that goes beyond that to which they originally agreed. There are several ways to do this.[16] If investigators wish to use samples for a research project beyond the scope of contributors' original consent, they could contact contributors and use samples only from those who give their specific consent to have their samples used for this additional project. Relative to current practice, adopting this way of proceeding would presumably in some ways be disadvantageous to researchers. For example, it would cost them time and money. And they might have fewer samples at their disposal, resulting in slower progress towards their research goals. One alternative would be for investigators to adopt a "tiered consent" approach:

At the time samples are collected, research participants are presented with a menu of options from which to choose, which may include general permission for future use, consent only for future uses related to the original study topic, consent for future uses unrelated to the original study topic, and specification that the investigators must obtain specific consent for any future use that differs from the original study.[17]

This approach might to some extent mitigate the disadvantages associated with specific consent. For it would presumably obviate the necessity of re-contacting each sample contributor for consent to a new research project, resulting in lower costs and more samples than the specific consent approach. But both approaches might reduce the pace of research that leads to the development of medicines and techniques that save lives and reduce suffering. That is a price that must be paid if we are to adopt research regulations that do more than those we currently have to prevent investigators from treating research participants merely as means.

I mentioned above that some philosophers would presumably reject something we have assumed here, namely a wide notion of using another, according to which in using a biological sample one is using a person. According to these philosophers, the Hybrid Account implies nothing regarding the moral permissibility of the researchers' behavior in our cases. However, moral problems analogous to the ones we have found in their behavior would emerge from applying a principle closely related to the Hybrid Account, namely what we might call the Part-Using Principle: Suppose an agent uses parts of a person's body. The agent acts (*pro tanto*) wrongly if it is reasonable for the agent to believe neither that the person can consent to her use of these parts, nor that he can share her end in using them. An application of this principle would, for example, yield the conclusion that the researchers who used the Havasupai blood for a study in evolutionary genetics acted (*pro tanto*) wrongly. Although I will not try to do so here, I believe that one might argue plausibly that an agent who uses parts of another person's body and fulfills the conditions specified in the Part-Using Principle would be failing to honor the other's dignity.

---

[16] This paragraph follows Mello and Wolf, "The Havasupai Indian Tribe Case," 204–5.
[17] ibid. 205.

8.4 Drug Trials and the Poor Abroad

Under what conditions is it morally permissible for US companies to engage in biomedical and clinical research in resource-limited countries? Many conditions are relatively uncontroversial. For example, competent adult subjects must give their voluntary, informed consent to participate in the research, and it must have scientific validity. But the moral permissibility of studies that fulfill such widely accepted conditions is nevertheless subject to lively and ongoing debate.

One sort of case that physicians, bioethicists, and lawyers have found particularly ethically challenging is illustrated in the following example. For the sake of clarity and economy of expression, the case is purely hypothetical. But it incorporates some features of widely discussed placebo-controlled trials.[18] A pharmaceutical company is conducting a placebo-controlled trial of a drug in a developing country that has very limited medical resources. American Institutional Review Boards (IRBs) would not allow such a trial to go forward in the United States. For in the United States there are readily available, proven drugs that treat the very same life-threatening medical condition that the new drug was designed to treat. The IRBs would demand that the effectiveness of the new drug be tested not against placebo, but against the current standard of care in an active-controlled trial. The drug company regards prospective subjects' informed, voluntary consent to participate to be a necessary condition of its using them in the trial. And all of them are competent and give such consent. They all understand that there is a 50 per cent chance that they will receive a placebo and a 50 per cent chance that they will receive the trial drug. If they do not take part in the study, they will get no effective treatment; if they do participate there is a considerable chance that they will receive such treatment. Taking part in the trial benefits all of the subjects in the sense that for each one it significantly raises the probability that he or she will get life-saving treatment. Conducting the research in the resource-limited country also benefits the international company. For it yields better data at lower cost than an active-controlled trial of the sort it would be permitted to conduct in the United States. In sum, both the study participants and the drug company stand to gain from the research.

In its most recent version, the Declaration of Helsinki seems to condemn such a study. It states that "the benefits, risks, burdens and effectiveness of a new intervention must be tested against those of the best current proven intervention."[19] The Declaration goes on to recognize two sets of circumstances in which placebo-controlled

---

[18]   See, for example, Peter Lurie and Sidney Wolfe, "Unethical Trials of Interventions to Reduce Perinatal Transmission of the Human Immunodeficiency Virus in Developing Countries," *The New England Journal of Medicine* 337 (1997) and Jennifer Hawkins, "Case Studies: The Havrix Trial and the Surfaxin Trial," in *Exploitation and Developing Countries*, ed. Jennifer Hawkins and Ezekiel Emanuel (Princeton: Princeton University Press, 2008).

[19]   World Medical Association, "Declaration of Helsinki. Ethical Principles for Medical Research Involving Human Subjects," *Journal of the Indian Medical Association* 107 (2009), Paragraph 32.

interventions are acceptable: if no current proven intervention exists or if placebo is necessary to evaluate the safety or efficacy of the intervention and the patients who get placebo "will not be subject to any risk of serious or irreversible harm."[20] Presumably a study participant would count as being subject to the risk of serious harm if he was given placebo instead of a currently available and highly effective treatment against a life-threatening condition, even though the placebo itself did not in any way exacerbate his condition. In any case, let us suppose that these exceptions do not apply in the hypothetical study described above: there is an effective treatment for the life-threatening condition and a placebo-controlled trial is not scientifically or methodologically necessary to evaluate the new treatment, although it would save the company money.

That the Declaration of Helsinki would condemn such a study does not, of course, entail that carrying one out would be wrong. Perhaps the Declaration is misguided. But one might think that there is a basis for holding that such a study would be morally impermissible to be found in the Kantian normative principles developed in Part I. Under what circumstances, if any, would the company be treating the study participants merely as means? When, if ever, would it be failing to respect their dignity as persons?

Dignity-deflationists charge that any ethical weight a notion of dignity has derives solely from its incorporating a familiar idea of respect for persons, according to which, in medical contexts, respect for persons requires getting their voluntary, informed consent before treating them or using them in research, as well as protecting their confidentiality. In a similar vein, Alan Wertheimer has, as we noted (7.1), recently claimed that it is not clear whether or how one agent (e.g., a drug company) who uses another (e.g., a subject in a drug trial) in a mutually beneficial and consensual interaction can in Kantian terms fail to respect the dignity of the other if, as in our hypothetical case, the agent has made securing the other's informed, voluntary consent to the transaction a necessary condition for going through with it. One aim of the upcoming sections is to shed light on how a drug company might, in such a scenario, be failing to respect the dignity of its research subjects.[21] Showing that will, in effect, show that the idea of respect for the dignity of persons developed in this book does not reduce to the idea of respect for persons commonly relied on in bioethics.

8.5 Drug Trials and Induced Vulnerability

We have found that one way to fail to respect the dignity of another is to use him merely as a means. But the Hybrid Account does not yield the conclusion that the pharmaceutical researchers are doing that. It is reasonable for them to believe that they have given the subjects an opportunity to opt out of being used in the study by

---

[20] World Medical Association, "Declaration of Helsinki," Paragraph 32.
[21] Wertheimer, *Rethinking the Ethics of Clinical Research*, 293.

withholding their agreement to participate. One might object to this conclusion, arguing that the researchers would surely realize that the subjects' desperate circumstances left them with no choice but to participate, and thus no opportunity to opt out of being used in the study. We have encountered this sort of claim before (4.1) and have found it implausible. That the effective promotion of a person's interests requires him to agree to being used in some way fails to entail that he cannot voluntarily consent (or dissent) to being used in this way. In any case, even if it was not reasonable for the researchers to believe that the subjects could consent to their use of them, it was reasonable for them to hold that the subjects could share the end they were pursuing in using them. More specifically, it would be reasonable for the researchers to believe that in willing their end of obtaining data on the effectiveness of the drug being tested, the subjects would not be willing to be thwarted in attaining their ends, including that of having their lives extended. For participating in the study would increase the likelihood that they would realize this aim.

But the Hybrid Account does not represent the only sufficient condition we have explored for someone's treating another merely as a means. In Chapter 4, we developed the Induced Vulnerability Account:

Suppose an agent uses another. She uses him merely as a means if:

1. It is reasonable for the agent to believe that something she has done to the other has contributed to his being in the position that unless the agent herself uses him, his well-being will diminish significantly.
2. It was foreseeable to the agent that she would contribute to his being in this position as well as avoidable that she would do so.
3. The agent was, as a practical matter, able to but did not give the other an opportunity to dissent from taking the risk that she (the agent) would contribute to the other's being in this position.
4. The end of the agent's use of the other is not limited to her discharging what she reasonably believes to be a moral obligation towards him.

For current purposes, the agent we have in mind is a pharmaceutical company, and the others, whom the agent uses, are people who enroll in the company's placebo-controlled trial in order to gain a significant chance of getting effective treatment for a life-threatening malady.

Under what circumstances would the Induced Vulnerability Account yield the conclusion that, in the sort of case we described, a pharmaceutical company was treating subjects merely as means? Circumstances in which the last condition is fulfilled are not at all hard to envisage. For even when pharmaceutical companies conduct trials that aim in part to fulfill moral obligations, such as that of helping those in need, they also have other aims such as developing profitable drugs.

The circumstances in which a company would meet the other conditions warrant more discussion. Regarding the first condition, some action of the company must contribute to the subjects' being unable to get effective treatment for their malady unless it uses them in its research. The company would, for example, have to have contributed to the unavailability to them (other than through its study) of effective medicine for the subjects' condition. It might have done this by lobbying for the current intellectual property regime, according to which pharmaceutical companies can get roughly twenty-year patents on medications they develop.[22] Their having these patents helps them to recuperate the costs of drug development and to make handsome profits. But it also helps to prevent the dissemination of generic versions of their drugs that would be affordable to patients in developing countries.

Second, it would have to be foreseeable to the company that its activities would contribute to the subjects' or, presumably, subjects like them, relying on it for effective treatment as well as avoidable that they would end up relying on it. This condition would be fulfilled if the company was aware that its lobbying efforts on behalf of the patent regime would prevent the dissemination of generic versions of drugs—versions that would be useful to people in developing countries. Moreover, the company would need to have the power to refrain from participating in such lobbying efforts. It is not hard to imagine these conditions being fulfilled. Indeed there is reason to believe that they have been.[23]

The third condition would be realized if a pharmaceutical company was able to but did not give citizens of developing countries the power to prevent it from foreseeably contributing to their possibly being dependent on being used by the company in order to get effective treatment for a life-threatening medical condition. The company would be able to give them this power if it could find out from them or, perhaps, from an official belonging to their democratically elected government, whether they would embrace the patent regime the company wished to promote and, if it turned out that they would not, refraining from doing any studies on this population. How often pharmaceutical companies have been able to do this and yet not done it is a complex empirical question that I will not try to address here. But some large pharmaceutical company may well have fulfilled condition 3. At the time it was lobbying for a twenty-year patent on newly developed drugs, it might have had plans to conduct a study in a particular country in which it ended up conducting one, and yet done nothing to find out whether the people of that country (as opposed, say, to its dictator) supported the patent regime the company was championing.[24]

In short, there are circumstances in which a pharmaceutical company would, in conducting the sort of placebo-controlled trial we described, be treating its experi-

---

[22] See Thomas Pogge, "Testing Our Drugs on the Poor Abroad," in *Exploitation and Developing Countries*, ed. Jennifer Hawkins and Ezekiel Emanuel (Princeton: Princeton University Press, 2008), 124–6.

[23] See Pogge, "Testing Our Drugs on the Poor Abroad," 124–6.

[24] It is worth noting that a pharmaceutical company might, according to the Induced Vulnerability Account, use research subjects in developing countries merely as means regardless of whether it used them for a placebo-controlled or an active-controlled study. Conditions 1–3 could conceivably be fulfilled in either instance.

mental subjects merely as means. But it is a complex empirical question how often (if ever) they do.

8.6 Honoring Dignity and Drug Trials on the Poor Abroad

If a pharmaceutical company uses its research subjects merely as means, then it thereby fails to respect their dignity, according to the Kant-Inspired Account of Dignity (KID) developed in Chapter 5. It fails to respect the subjects' dignity, even if it evinces respect for persons, according to a familiar idea of such respect, by gaining their voluntary, informed consent before proceeding and honoring their confidentiality.

But a company can fail to respect the dignity of its research subjects even if it is not treating them merely as a means, according to KID. The part of the account relevant to our discussion is as follows:

The dignity of persons is a special status that they possess by virtue of having the capacities constitutive of personhood. This status has several features, including the following:

1. It is such that a person ought not to use another merely as a means. This first aspect of persons' special status is lexically prior to all of those that follow.
2. All persons have a status such that if an agent treats them in some way, then she ought to treat them as having unconditional, transcendent value.
3. The status of persons is such that if an agent treats another in some way, she ought to treat him as having an unconditional, transcendent value that does not change as a result of what the other does or of the agent's relation to him, apart from the following exceptions . . .

The account goes on to specify exceptions to 3 that do not concern us here. KID holds that an agent's treatment of a person respects his dignity only if it accords with the special status just described. The pharmaceutical company's treatment of its subjects might fail to respect their dignity as a result of its running afoul of the notion that all persons have a status such that if an agent treats them in some way, it ought to treat them as having unconditional, transcendent value.

Let me explain. For a person to have this value is for her to have (positive) value in every possible context in which she exists: a value that has no equal in any amount of anything that is not a person. Someone treats another as having unconditional, transcendent value only if, in the given context, the action she performs is among those that someone might perform if he reasonably believed his action to be (successfully and absolutely) constrained by his holding them to have this value. Now suppose that an executive at the pharmaceutical company is free to choose between conducting an active-controlled or placebo-controlled trial in the developing country. Potential subjects, through their community representatives, have requested that the company do the former kind of trial, but have also indicated that they would welcome the latter. In order to maximize the company's profits, the executive goes ahead with the less expensive, placebo-controlled trial, even though she believes reasonably that conducting an active-controlled trial would preserve more lives overall. In so doing, she treats the trial subjects in some way. But her action is not among those that someone might in this context perform if he reasonably believed his action to be (successfully and absolutely) constrained by his holding persons to have unconditional, transcendent value. If one takes one's treatment of another to be constrained by holding persons to have this special value, then in these circumstances one would take the course of action that would preserve more persons. In other words, one would conduct the active-controlled as opposed to the placebo-controlled trial. To refrain from conducting the active-controlled trial is to privilege something of mere price, namely profits, over beings of unconditional, transcendent value, namely persons. The executive is thus failing to respect the dignity of the persons in the placebo arm of the trial: the persons who do not receive, but could have received, life-saving treatment as a part of a company-sponsored study.

That the executive's action would fail to respect the dignity of some persons fails to entail that her action is morally wrong, all things considered. We found in Chapter 5 that, according to many of us, some actions that fail to respect the dignity of persons are morally permissible. For example, many of us believe that it would be morally permissible for someone to use another merely as a means if he reasonably held this to be the only way to save thousands, and that a person would not act contrary to morality if he paid for hospice care for his terminally ill child instead of sending this money abroad to save the life of a stranger with whom his only contact had been that of receiving from him a credible request for aid. However, KID implies that these actions fail to honor the dignity of persons.

Is the executive's action morally permissible, all things considered? Her choice to do the placebo-controlled trial might save her company enough money to make possible larger bonuses for its employees, extra money they might use to enhance the well-being of themselves and their loved ones. Moreover, the subjects who enroll in the placebo-controlled trial benefit by augmenting their chances of survival. So we have some reason to judge that despite its violation of dignity, the executive's carrying out the trial is morally permissible.

But it is doubtful whether we have sufficient reason to believe this. In Chapter 5 we made some conjectures regarding conditions under which we judge a person to fail to respect someone's dignity and yet not to act wrongly, all things considered. According to one conjecture, if an agent is using another and failing to respect his dignity, then we need stronger reason to conclude that his failure to respect the other's dignity is morally permissible than we would need if she were not using the other (5.6). The executive is, of course, using the study subjects who are assigned to the placebo arm of the trial. And the benefit she seeks in conducting such a trial, rather than an active-controlled trial, is greater profit for her company. This benefit does not seem to warrant a violation of the subjects' dignity.

In sketching this example, we assumed that the pharmaceutical company executive had the authority to opt either for a placebo-controlled or for an active-controlled trial in a developing country. But suppose that she had instead to choose between a placebo-controlled trial in a developing country and an active-controlled trial in the United States. Her company is a start-up, and conducting an active-controlled trial in a developing country would literally be financially infeasible. The company contacted medical authorities and patient groups in the developing country who indicated a preference for an active-controlled trial, but said that if one were not possible, they would like to host a placebo-controlled trial. In this case, it would be consistent with KID for the executive to go through with the placebo-controlled trial. This action would be among those someone in this context might perform if he reasonably believed his action to be (successfully and absolutely) constrained by his holding persons to have unconditional, transcendent value. It would be the action that, in the context, would preserve as many persons as possible. If there were no study for them to join, those in the developing country would receive no treatment, while those in the United States would nevertheless receive effective treatment. Of course, whether an action honors persons' dignity according to KID depends in part on the range of actions open to the agent.

But what options are really open to pharmaceutical companies in terms of the trials they conduct? This is an empirical question that I will not try to answer here. But in light of their typically expansive profit margins, it is hard to believe that they cannot afford to conduct trials in developing countries rather than developed ones, thereby saving lives and reducing suffering.[25]

In any case, KID has enabled us to meet Wertheimer's challenge. We have isolated conditions under which an agent's use of another amounts to a failure on her part to respect the other's dignity, even though this use is mutually beneficial, the other has given his informed, voluntary consent to it, and the agent has made it a condition of her proceeding with the use that the other give such consent. Pharmaceutical companies' use of research subjects can fail to respect their dignity and thereby be wrong even if it takes place against the background of both parties agreeing to it and expecting to benefit from it.

Moreover, we have shown that the ethical content of the demand to respect the dignity of persons is not exhausted by the content of a common notion of respect for persons, according to which doing that requires that one obtain the voluntary,

[25] For an indication of drug company profits, see Marcia Angell, *The Truth About the Drug Companies* (New York: Random House, 2004), 3. According to *Fortune Magazine*, among the 50 most profitable companies in the United States during 2010 were six pharmaceutical companies. See "Annual Ranking of America's Largest Corporations: Top Companies: Most Profitable," *Fortune* <http://money.cnn.com/magazines/fortune/fortune500/2010/performers/companies/profits/>.

informed consent of research subjects and protect their confidentiality. In some of the cases that have been our focus, pharmaceutical companies *do* respect persons according to this common notion, but, according to those of us who find KID plausible, they fail to respect the dignity of persons and thereby act wrongly.

8.7 Concluding Remarks

This book has, I hope, shown that a Kantian account of the dignity of persons (KID) can illuminate moral questions that arise concerning medical research, markets in organs, and the distribution of scarce, life-saving resources. KID asserts that two notions are central to honoring the dignity of persons: that we ought not to treat others merely as means and that we ought to treat them as having unconditional, transcendent value.

On the book's construal of treating others merely as means, it makes sense to say that in doing so one is acting (*pro tanto*) wrongly. In just using another, a person is not merely doing something that she ought to see as wrong. She is also (typically) doing something that *is* wrong. In treating another merely as a means, a person is not just indicating a flaw in her character, but (typically) engaging in morally impermissible conduct. An agent can use another merely as a means and thereby act (*pro tanto*) wrongly, we have found, even if the other has given his voluntary, informed consent to her use of him. That another has agreed to an agent's using him, even on the basis of an understanding of what the use will consist in, to what end, and with what effect on him, is consistent with her doing wrong in using him.

We have specified what treating another merely as a means consists in, at least according to one construal of common sense. But we have not focused on the question of why doing so is *pro tanto* wrong. Regarding the Hybrid Account, we might say the following: For an agent to treat another merely as a means is (typically) for her to fail to take proper account of the other's agency in her use of him. And that is what makes it wrong. In order for an agent to take proper account of another's agency, it must (typically) at least be reasonable for the agent to believe that the other can share her (proximate) end in using him or that he can avert this use by withholding his agreement to it. This remark is, in effect, a statement of what is implicit in the Hybrid Account itself. As soon as we pinpoint what treating another merely as a means amounts to, we recognize that it is generally wrong. For most of us, I venture, no further consideration is needed or likely to be helpful.

KID stipulates not only that honoring persons' dignity requires refraining from treating others merely as means, but also that it demands treating persons as having unconditional, transcendent value. KID thereby captures some of the attractive features of a traditional Kantian account. According to both, we fail to respect someone's dignity if we treat him as exchangeable for something of mere price, such as precious gems or stock options. But unlike a traditional Kantian account, KID is free from the implausible implication that in making trade-offs between persons we always fail to honor someone's dignity. It is consistent with respecting the dignity of persons, according to KID, for someone to intentionally sacrifice his own life in order to preserve what he takes to be the greater value inherent in the lives of several others, or for someone whose job it is to allocate scarce, life-saving resources to choose to save a younger rather than an older person on the grounds that the younger person will thrive for a lot longer.

KID might be more promising than a traditional Kantian account of the dignity of persons. But do we really need such an intricate account—especially when, as we acknowledge, it is not complete? Deflationists argue that when discourse on dignity is denuded of grandiloquent rhetoric, what remains is a simple idea. In order to respect the dignity of persons, we need to refrain from interfering with the choices of autonomous individuals, unless these choices harm others. If, before using another person, an agent gets the other's voluntary, informed consent to it, then her use of him does not interfere with his choices, and so respects his dignity as a person according to this view.

But along with simplicity comes an inability to discern important features of the moral landscape, we have found. The deflationist view overlooks serious failures to honor the dignity of persons. For example, it does not recognize cases in which an agent treats another merely as a means even though the other has given his voluntary, informed consent to this treatment. Moreover, it overlooks that when we use others, respecting their dignity can require us to do more than is in our interest in order to preserve their lives, even when our use of the others benefits them, and they have freely agreed to it.

The light KID shines on actions that, upon reflection, we find to be morally problematic underscores its usefulness. This book has, I hope, shown KID to be a worthy representative of Kantianism in normative ethics. But the process of justifying it is ongoing. For example, its implications need to be explored in greater detail in areas, such as resource allocation, which have been dominated by utilitarian thinking. And KID's plausibility must be compared to that of rival principles that stem from outside of the Kantian tradition.

Bibliography

Abadie, Alberto, and Sebastien Gay. "The Impact of Presumed Consent Legislation on Cadaveric Organ Donation: A Cross-Country Study." *Journal of Health Economics* 25 (2006): 599–620.

Almashat, Sammy, Brian Ayotte, Barry Edelstein, et al. "Framing Effect Debiasing in Medical Decision Making." *Patient Education and Counseling* 71 (2007): 102–7.

American Psychiatric Association. *Diagnostic and Statistical Manual of Mental Disorders DSM-IV-TR Fourth Edition*. Washington, D.C.: American Psychiatric Association, 2000.

Angell, Marcia. *The Truth About the Drug Companies*. New York: Random House, 2004.

Annas, George, and Michael Grodin. "The Nuremberg Code." In *The Oxford Textbook of Clinical Research Ethics*, edited by Ezekiel Emanuel et al., 136–40. Oxford: Oxford University Press, 2008.

"Annual Ranking of America's Largest Corporations: Top Companies: Most Profitable." *Fortune* <http://money.cnn.com/magazines/fortune/fortune500/2010/performers/companies/profits/>. Accessed August 11, 2012.

Arlington National Cemetery "Ross Andrew McGinnis: Specialist, United States Army" <http://www.arlingtoncemetery.net/ramcginnis.htm>. Accessed August 11, 2012.

Armstrong, Katrina, J. Schwartz, Genevieve Fitzgerald, et al. "Effect of Framing as Gain Versus Loss on Understanding and Hypothetical Treatment Choices: Survival and Mortality Curves." *Medical Decision Making* 22 (2002): 76–83.

Arnesen, Trude, and Mari Trommald. "Are QALYs Based on Time Trade-Off Comparable: A Systematic Review of TTO Methodologies." *Health Economics* 14 (2005): 39–53.

Baird, Robert, and Stuart Rosenbaum. *Pornography: Private Right or Public Menace?* Revised edition. Amherst, NY: Prometheus Books, 1998.

Baron, Marcia. "Manipulativeness." *Proceedings and Addresses of the American Philosophical Association* 77 (2003): 37–54.

Beauchamp, Tom. "Autonomy and Consent." In *The Ethics of Consent: Theory and Practice*, edited by Franklin Miller and Alan Wertheimer, 55–78. Oxford: Oxford University Press, 2010.

—— and James Childress. *Principles of Biomedical Ethics*, 6th edition. New York: Oxford University Press, 2009.

Berker, Selim. "The Normative Insignificance of Neuroscience." *Philosophy and Public Affairs* 37 (2009): 293–329.

Bittner, Rüdiger. *Doing Things for Reasons*. Oxford: Oxford University Press, 2001.

Blackwell, Mark. "'Extraneous Bodies': The Contagion of Live-Tooth Transplantation in Late-Eighteenth-Century England." *Eighteenth-Century Life* 28 (2004): 21–68.

Bognar, Greg. "Age-Weighting." *Economics and Philosophy* 24 (2008): 167–89.

—— and Samuel Kerstein. "Saving Lives and Respecting Persons," *Journal of Ethics and Social Philosophy* 5(2) (2010): 1–20.

Bramstedt, K. A., and J. Xu. "Checklist: Passport, Plane Ticket, Organ Transplant." *American Journal of Transplantation* 7 (2007): 1698–701.

Brock, Dan. "Voluntary Active Euthanasia." *Hastings Center Report* 22 (1992): 10–22.

——. *Life and Death: Philosophical Essays in Biomedical Ethics*. Cambridge: Cambridge University Press, 1993.

Candilis, Philip, and Charles Lidz. "Advances in Informed Consent Research." In *The Ethics of Consent*, edited by Franklin Miller and Alan Wertheimer. Oxford: Oxford University Press, 2010.

Caplan, Arthur, Beatriz Domínguez-Gil, Rafael Matesanz, and Carmen Prior. "Trafficking in Organs, Tissues and Cells and Trafficking in Human Beings for the Purpose of the Removal of Organs." Strasbourg: Joint Council of Europe/United Nations Study, 2009.

Card, Claudia. *Confronting Evils: Terrorism, Torture, Genocide*. Cambridge: Cambridge University Press, 2010.

Cherry, Mark. *Kidney for Sale by Owner: Human Organs, Transplantation, and the Market*. Washington, D.C.: Georgetown University Press, 2005.

Chochinov, H. et al. "Dignity in the Terminally Ill: A Cross-Sectional, Cohort Study." *Lancet* 360 (2002): 2026–30.

Christiano, Thomas. "Two Conceptions of the Dignity of Persons." *Jahrbuch für Recht und Ethik* 16 (2008): 101–26.

Cico, Stephen, Eva Vogeley, and William Doyle. "Informed Consent Language and Parents' Willingness to Enroll Their Children in Research." *IRB: Ethics and Human Research* 33 (2011): 6–13.

Cohen, Cynthia. "Public Policy and the Sale of Human Organs." *Kennedy Institute of Ethics Journal* 12 (2002): 47–64.

Cohen, Lawrence. "Where It Hurts: Indian Material for an Ethics of Organ Transplantation." *Daedalus* 128 (1999): 135–65.

Committee on Increasing Rates of Organ Donation. *Organ Donation: Opportunities for Action*. Washington, D.C.: The National Academies Press, 2006.

Conde, Carlos. "Philippines Bans Kidney Transplants for Foreigners." *The New York Times*, April 30, 2008.

Cummiskey, David. *Kantian Consequentialism*. New York: Oxford University Press, 1996.

——. "Dignity, Contractualism, and Consequentialism." *Utilitas* 20 (2008): 383–408.

Daniels, Norman. "Wide Reflective Equilibrium and Theory Acceptance in Ethics." *Journal of Philosophy* 76 (1979): 256–82.

——. *Just Health: Meeting Health Needs Fairly*. Cambridge: Cambridge University Press, 2008.

—— Susannah Rose, and Ellen Zide. "Disability, Adaptation, and Inclusion." In *Disability and Disadvantage*, edited by Kimberley Brownlee and Adam Cureton, 54–85. Oxford: Oxford University Press, 2009.

Davis, F. Daniel, and President's Council on Bioethics (US). "Human Dignity and Respect for Persons: A Historical Perspective on Public Bioethics." In *Human Dignity and Bioethics: Essays Commissioned by the President's Council on Bioethics*, 19–36. Washington, D.C.: President's Council on Bioethics, 2008.

Davis, Nancy. "Using Persons and Common Sense." *Ethics* 94 (1984): 387–406.

Dean, Richard. *The Value of Humanity in Kant's Moral Theory*. Oxford: Clarendon Press, 2006.

——. "Does Neuroscience Undermine Deontological Theory?" *Neuroethics* 3 (2009): 43–60.

DeGrazia, David. *Taking Animals Seriously: Mental Life and Moral Status*. Cambridge: Cambridge University Press, 1996.

——. *Human Identity and Bioethics*. Cambridge: Cambridge University Press, 2005.

Denis, Lara. "Kant's Conception of Duties Regarding Animals: Reconstruction and Reconsideration." *History of Philosophy Quarterly* 17 (2000): 405–23.

Diagnostic and Statistical Manual of Mental Disorders, DSM-IV-TR. Fourth Edition. "Major Depressive Disorder" (2000).

Duff, R. A. *Trials and Punishments.* Cambridge: Cambridge University Press, 1986.

Dworkin, Gerald. *The Theory and Practice of Autonomy.* Cambridge: Cambridge University Press, 1988.

Emanuel, Ezekiel et al. *The Oxford Textbook of Clinical Research Ethics.* Oxford: Oxford University Press, 2008.

——and Jerry Menikoff. "Reforming the Regulations Governing Research with Human Subjects." *New England Journal of Medicine* 365 (2011): 1145–50.

Fanselow, Ryan. "Moral Intuitions and Their Role in Justification." PhD Dissertation. University of Maryland, College Park, 2011.

Frankfurt, Harry. "Freedom of the Will and the Concept of a Person." *Journal of Philosophy* 68 (1971): 5–20.

Garcia-Retamero, Racio, and Mirta Galesic. "How to Reduce the Effect of Framing on Messages About Health." *Journal of General Internal Medicine* 25 (2010): 1323–9.

Gigerenzer, Gerd, Wolfgang Gaissmaier, Elke Kurz-Milcke, et al. "Helping Doctors and Patients Make Sense of Health Statistics." *Psychological Science in the Public Interest* 8 (2008): 53–96.

Gill, Michael. "Presumed Consent, Autonomy, and Organ Donation." *Journal of Medicine and Philosophy* 29 (2004): 37–59.

——and Robert Sade. "Paying for Kidneys: The Case against Prohibition." *Kennedy Institute of Ethics Journal* 12 (2002): 17–45.

Glover, Jonathan. *Choosing Children: Genes, Disability, and Design.* Oxford: Oxford University Press, 2006.

Goyal, Madhav, et al. "Economic and Health Consequences of Selling a Kidney in India." *Journal of the American Medical Association* 288 (2002): 1589–93.

Harman, Gilbert. *The Nature of Morality.* Oxford: Oxford University Press, 1977.

Harris, John. *The Value of Life: An Introduction to Medical Ethics.* London: Routledge & Kegan Paul, 1985.

——. "QALYfying the Value of Life." *Journal of Medical Ethics* 13 (1987): 117–23.

Haywood, H. L. "Rotary Ethics." *The Rotarian*, 1918.

Hawkins, Jennifer. "Case Studies: The Havrix Trial and the Surfaxin Trial." In *Exploitation and Developing Countries*, edited by Jennifer Hawkins and Ezekiel Emanuel, 53–60. Princeton: Princeton University Press, 2008.

Healy, Kieran. "Do Presumed Consent Laws Raise Organ Procurement Rates?" *DePaul Law Review* 55 (2006): 1017–43.

Herman, Barbara. *The Practice of Moral Judgment.* Cambridge, Mass.: Harvard University Press, 1993.

Hill, Thomas E., Jr. *Dignity and Practical Reason in Kant's Moral Theory.* Ithaca: Cornell University Press, 1992.

——. "Respect for Humanity." *The Tanner Lectures on Human Values* (1997) <http://www.tannerlectures.utah.edu/lectures/documents/Hill97.pdf>. Accessed August 11, 2012.

——. *Respect, Pluralism, and Justice: Kantian Perspectives.* Oxford: Oxford University Press, 2000.

——. *Human Welfare and Moral Worth: Kantian Perspectives.* Oxford: Clarendon Press, 2002.

——. "Treating Criminals as Ends in Themselves." *Jahrbuch für Recht und Ethik* 11 (2003): 17–36.

Hippen, Benjamin. "In Defense of a Regulated Market in Kidneys from Living Vendors." *Journal of Medicine and Philosophy* 30 (2005): 593–626.

——. "Organ Sales and Moral Travails: Lessons from the Living Kidney Vendor Program in Iran." *Policy Analysis* 614 (2008): 1–17.

Hirose, Iwao. "Moral Aggregation." (2010), monograph.

Ivanovski, N. et al. "The Outcome of Commercial Kidney Transplant Tourism in Pakistan." *Clinical Transplantation* 25 (2011): 171–3.

Kamm, Frances. *Morality, Mortality, Volume 1: Death and Whom to Save from It.* New York: Oxford University Press, 1993.

——. *Morality, Mortality, Volume 2: Rights, Duties, and Status.* New York: Oxford University Press, 1996.

——. "A Right to Choose Death." *Boston Review* Summer (1997), section V.

——. *Intricate Ethics: Rights, Responsibilities, and Permissible Harm.* Oxford: Oxford University Press, 2007.

Kant, Immanuel. *Groundwork of the Metaphysics of Morals.* Translated by Mary Gregor. In *Immanuel Kant: Practical Philosophy.* Cambridge: Cambridge University Press, 1996.

——. *The Critique of Practical Reason.* Translated by Mary Gregor. In *Immanuel Kant: Practical Philosophy.* Cambridge: Cambridge University Press, 1996.

——. *The Metaphysics of Morals.* Translated by Mary Gregor. In *Immanuel Kant: Practical Philosophy.* Cambridge: Cambridge University Press, 1996.

——. *Lectures on Ethics.* Translated by Peter Heath. Cambridge: Cambridge University Press, 1997.

Kaufmann, Paulus. "Instrumentalization: What Does It Mean to Use a Person?" In *Humiliation, Degradation, Dehumanization: Human Dignity Violated,* edited by Paulus Kaufmann, Hannes Kuch, Christian Neuhäuser and Elaine Webster, 57–65. Dordrecht: Springer, 2011.

Kerstein, Samuel. "Korsgaard's Kantian Arguments for the Value of Humanity." *Canadian Journal of Philosophy* 31 (2001): 23–52.

——. *Kant's Search for the Supreme Principle of Morality.* Cambridge: Cambridge University Press, 2002.

——. "Autonomy and Practical Law." *Philosophical Books* 49 (2008): 107–13.

——. Rev. of *Kantian Ethics* by Allen Wood. *Ethics* 118 (2008): 761–7.

——. "Treating Oneself Merely as a Means." In *Kant's Ethics of Virtues,* edited by Monika Betzler, 201–18. Berlin: Walter de Gruyter, 2008.

——. "Autonomy, Moral Constraints, and Markets in Kidneys." *Journal of Medicine and Philosophy* 34 (2009): 573–85.

——. "Kantian Condemnation of Commerce in Organs." *Kennedy Institute of Ethics Journal* 19 (2009): 147–69.

——"Death, Dignity, and Respect," *Social Theory and Practice* 35 (2009), 505–530.

——. "Treating Others Merely as Means." *Utilitas* 21 (2009): 163–80.

——. "Treating Consenting Adults Merely as Means," *Oxford Studies in Normative Ethics* 1 (2011): 51–74.

——. "Imperatives: Categorical and Hypothetical." In *International Encyclopedia of Ethics,* edited by Hugh LaFollette. Oxford: Blackwell, 2012.

——— and Greg Bognar. "Complete Lives in the Balance." *American Journal of Bioethics* 10 (2010): 37–45.

———. "Response to Open Peer Commentaries On 'Complete Lives in the Balance.'" *American Journal of Bioethics* 10 (2010): W3–W5.

Korsgaard, Christine. *Creating the Kingdom of Ends*. Cambridge: Cambridge University Press, 1996.

———. *The Sources of Normativity*. Cambridge: Cambridge University Press, 1996.

Langbein, John. "On the Myth of Written Constitutions: The Disappearance of Criminal Jury Trial." *Harvard Journal of Law and Public Policy* 15 (1993): 119–27.

Levine, Robert. "Respect for Children as Research Subjects." In *Lewis's Child and Adolescent Psychiatry*, edited by Andres Martin and Fred Volkmar, 140–9. Philadelphia: Lippincott, Williams, and Wilkins, 2007.

Longthorne, Anders, Rajesh Subramanian, and Chou-Lin Chen. *An Analysis of the Significant Decline in Motor Vehicle Traffic Crashes in 2008*. Edited by US Department of Transportation. Washington, D.C.: National Highway Traffic Safety Administration, 2010.

Luckey, John. "Flag Protection: A Brief History and Summary of Recent Supreme Court Decisions and Proposed Constitutional Amendment." In *Congressional Research Service Report for Congress*, 2005.

Lurie, Peter, and Sidney Wolfe. "Unethical Trials of Interventions to Reduce Perinatal Transmission of the Human Immunodeficiency Virus in Developing Countries." *The New England Journal of Medicine* 337 (1997): 1–5.

McCoy, Candace. *Politics and Plea Bargaining*. Philadelphia: University of Pennsylvania Press, 1993.

———. "Plea Bargaining as Coercion: The Trial Penalty and Plea Bargaining Reform." *Criminal Law Quarterly* 50 (2005): 67–107.

McGrath, Sarah, and Lidet Tilahun. *Hip Hop and Philosophy: Rhyme 2 Reason*, edited by Derrick Darby and Tommie Shelby. Chicago: Open Court, 2005.

Macklin, Ruth. *Surrogates and Other Mothers: The Debates over Assisted Reproduction*. Philadelphia: Temple University Press, 1994.

———. "Dignity Is a Useless Concept." *British Medical Journal* 327 (2003): 1419–20.

McMahan, Jeff. *The Ethics of Killing: Problems at the Margins of Life*. New York: Oxford University Press, 2002.

McNeil, Barbara, Stephen Pauker, Harold Sox, and Amos Tversky. "On the Elicitation of Preferences for Alternative Therapies." *New England Journal of Medicine* 306 (1982): 1259–62.

Malakoutian, T. et al. "Socioeconomic Status of Iranian Living Unrelated Kidney Donors: A Multicenter Study." *Transplantation Proceedings* 39 (2007): 824–5.

Manson, Neil, and Onora O'Neill. *Rethinking Informed Consent in Bioethics*. Cambridge: Cambridge University Press, 2007.

Markow, Therese, et al. "HLA Polymorphism in the Havasupai: Evidence for Balancing Selection." *American Journal of Human Genetics* 53 (1993): 943–52.

Matas, Arthur. "Why We Should Develop a Regulated System of Kidney Sales: A Call for Action." *Clinical Journal of the American Society of Nephrology* 1 (2006): 1129–32.

———Benjamin Hippen, and Sally Satel. "In Defense of a Regulated System of Compensation for Living Donation." *Current Opinion in Organ Transplantation* 13 (2008): 379–85.

"Medal of Honor Recipients: Iraq." US Army Center of Military History. <http://www.history.army.mil/html/moh/iraq.html>. Accessed August 11, 2012.

Mediavilla, Sam. "President Wants to Stop Organ Trafficking." *Manila Times*, February 8, 2007.

Mello, Michelle, and Leslie Wolf. "The Havasupai Indian Tribe Case—Lessons for Research Involving Stored Biologic Samples." *New England Journal of Medicine* 363 (2010): 204–7.

Menzel, Paul. "Complete Lives, Short Lives, and the Challenge of Legitimacy." *American Journal of Bioethics* 10 (2010): 50–2.

Miller, Franklin. "Consent to Clinical Research." In *The Ethics of Consent*, edited by Franklin Miller and Alan Wertheimer, 375–404. Oxford: Oxford University Press, 2010.

——and Alan Wertheimer. "Payment for Research Participation: A Coercive Offer?" *Journal of Medical Ethics* 34 (2008): 389–92.

Mills, Claudia. "Politics and Manipulation." *Social Theory and Practice* 21 (1995): 97–112.

Morelli, Mario. "Commerce in Organs: A Kantian Critique." *Journal of Social Philosophy* 30 (1999): 315–24.

Naqvi, Syed et al. "A Socioeconomic Survey of Kidney Vendors in Pakistan." *Transplant International* 20 (2007): 934–39.

——et al. "Health Status and Renal Function Evaluation of Kidney Vendors: A Report from Pakistan." *American Journal of Transplantation* 8 (2008): 1444–50.

Nejatisafa, Ali-Akbar et al. "Quality of Life and Life Events of Living Unrelated Kidney Donors in Iran: A Multicenter Study." *Transplantation* 86 (2008): 937–40.

Nord, Erik. *Cost-Value Analysis in Health Care: Making Sense out of QALYs*. Cambridge: Cambridge University Press, 1999.

Nozick, Robert. *Anarchy, State, and Utopia*. New York: Basic Books, 1974.

Nussbaum, Martha. *Frontiers of Justice: Disability, Nationality, Species Membership*. Cambridge, Mass.: Harvard University Press, 2006.

O'Neill, Onora. *Constructions of Reason*. Cambridge: Cambridge University Press, 1989.

——. "Autonomy: The Emperor's New Clothes." *Aristotelian Society Supplementary Volume* 77 (2003): 1–21.

Organ Procurement and Transplantation Network (OPTN) "Overall by Organ: Current U.S. Waiting List," 2011.

——. "Transplants in the U.S. By State," 2011.

Parfit, Derek. *On What Matters*, Vols I and II. Oxford: Oxford University Press, 2011.

Pew Forum on Religion and Public Life. "U.S. Religious Landscape Survey: Religious Beliefs and Practices: Diverse and Politically Relevant." Washington, D.C., June 2008.

Pinker, Steven. "The Stupidity of Dignity." *The New Republic* May 28 (2008). Available at <http://www.tnr.com>. Accessed August 11, 2012.

Pogge, Thomas. "Testing Our Drugs on the Poor Abroad." In *Exploitation and Developing Countries*, edited by Jennifer Hawkins and Ezekiel Emanuel, 105–41. Princeton: Princeton University Press, 2008.

Porter, Joan, and Greg Koski. "Regulations for the Protection of Humans in Research in the United States." In *The Oxford Textbook of Clinical Research Ethics*, edited by Ezekiel Emanuel et al., 156–67. Oxford: Oxford University Press, 2008.

Pulley, Jill et al. "Principles of Human Subjects Protections Applied in an Opt-Out, De-Identified Biobank." *Clinical and Translational Science* 3 (2010): 42–8.

Radcliffe Richards, Janet, A. Daar, and R. Guttman. "The Case for Allowing Kidney Sales." *Lancet* 351 (1998): 1950–2.

Rawls, John. *A Theory of Justice*. Cambridge, Mass.: Harvard University Press, 1971.

Reath, Andrews. *Agency and Autonomy in Kant's Moral Theory*. Oxford: Oxford University Press, 2006.

"Regulatory Changes in ANPRM." US Department of Health and Human Services. Available at <http://www.hhs.gov/ohrp/humansubjects/anprmchangetable.html>. Accessed August 13. 2012.

Richardson, Henry. "Discerning Subordination and Inviolability: A Comment on Kamm's *Intricate Ethics*." *Utilitas* 20 (2008): 81–91.

Robichaud, Christopher. "To Turn or Not to Turn: The Ethics of Making Vampires." In *True Blood and Philosophy: We Wanna Think Bad Things with You*, edited by George A. Dunn and Rebecca Housel, 7–18. Hoboken, N.J.: Wiley, 2010.

Rodin, David. *War and Self-Defense*. Oxford: Oxford University Press, 2002.

Rothstein, Mark. "Is Deidentification Sufficient to Protect Health Privacy in Research?" *American Journal of Bioethics* 10 (2010): 3–11.

Sataline, S., C. Bray, and G. Fields. "Four Men Held in U.S. Terror Case." *The Wall Street Journal*, May 22, 2009, A3.

Satz, Debra. *Why Some Things Should Not Be for Sale: the Moral Limits of Markets*. New York: Oxford University Press, 2010.

Scanlon, Thomas. *What We Owe to Each Other*. Cambridge, Mass.: Harvard University Press, 1998.

——. *Moral Dimensions: Permissibility, Meaning, Blame*. Cambridge, Mass.: Harvard University Press, 2008.

Schaber, Peter. "Using People Merely as a Means." Unpublished manuscript.

Schönecker, Dieter. "How Is a Categorical Imperative Possible?" In *Groundwork for the Metaphysics of Morals*, edited by Christoph Horn and Dieter Schönecker. Berlin: Walter de Gruyter, 2006.

Segev, Dorry, et al. "Perioperative Mortality and Long-Term Survival Following Live Kidney Donation." *The Journal of the American Medical Association* 303 (2010): 959–66.

Sensen, Oliver. "Dignity and the Formula of Humanity (Ad IV 429, IV 435)." In *Kant's "Groundwork of the Metaphysics of Morals": A Critical Guide*, edited by Jens Timmermann, 102–18. Cambridge: Cambridge University Press, 2010.

——. *Kant on Human Dignity*. Berlin: Walter de Gruyter, 2011.

Shimazono, Y. "The State of the International Organ Trade: A Provisional Picture Based on Integration of Available Information." *Bulletin of the World Health Organization* 85 (2007): 955–62.

Sidgwick, Henry. *The Methods of Ethics*, 7th ed. Indianapolis, Indiana: Hackett, 1981.

Singer, Peter. "Sidgwick and Reflective Equilibrium." *The Monist* 58 (1974): 490–517.

——. "Ethics and Intuitions." *The Journal of Ethics* 9 (2005): 331–52.

——et al. "Double Jeopardy and the Use of QALYs in Health Care Allocation." *Journal of Medical Ethics* 21 (1995): 144–50.

Taurek, John. "Should the Numbers Count?" *Philosophy and Public Affairs* 6 (1977): 293–316.

Taylor, James. *Stakes and Kidneys*. London: Ashgate, 2005.

Taylor, James. "Autonomy and Organ Sales Revisited." *Journal of Medicine and Philosophy* 34 (2009): 632–8.

Thomson, Judith Jarvis. *Rights, Restitution, and Risk: Essays in Moral Theory*. Cambridge, Mass.: Harvard University Press, 1986.

Thomson, Ross. "The Development of the Person: Social Understanding, Relationships, Conscience, Self." In *Handbook of Child Psychology: Social, Emotional, and Personality Development*, Volume 3, edited by Nancy Eisenberg, 26–98. Hoboken, N.J.: John Wiley & Sons, 2006.

Timmermann, Jens. "The Individualist Lottery: How People Count, but Not Their Numbers." *Analysis* 64 (2004): 106–12.

———. *Kant's Groundwork of the Metaphysics of Morals: A Critical Guide*. Cambridge: Cambridge University Press, 2009.

Tsuchiya, Takashi. "The Imperial Japanese Experiments in China." In *The Oxford Textbook of Clinical Research Ethics*, edited by Ezekiel Emanuel et al., 31–45. Oxford: Oxford University Press, 2008.

Ubel, Peter, George Loewenstein, and Christopher Jepson. "Whose Quality of Life? A Commentary Exploring Discrepancies between Health State Evaluations of Patients and the General Public." *Quality of Life Research* 12 (2003): 599–607.

Uniacke, Suzanne. *Permissible Killing: The Self-Defence Justification of Homicide*. Cambridge: Cambridge University Press, 1994.

US Department of Health and Human Services, Office for Human Research Protections (OHRP). "Guidance on Research Involving Coded Private Information or Biological Specimens," Washington, D.C., October 16, 2008.

Van der Graaf, Rieke, and Johannes Van Delden. "On Using People Merely as Means in Clinical Research." *Bioethics* 25 (2011): 76–83.

Velleman, J. David. "A Right of Self-Termination?" *Ethics* 109 (1999): 606–28.

Waldron, Jeremy. "From Authors to Copiers: Individual Rights and Social Values in Intellectual Property." In *Intellectual Property Rights: Critical Concepts in Law*, edited by David Vaver, 114–56. London: Routledge, 2006.

Walsh, Adrian, and Richard Giulianotti. *Ethics, Money, and Sport*. New York: Routledge, 2007.

Wasserman, David, and Alan Strudler. "Can a Nonconsequentialist Count Lives?" *Philosophy and Public Affairs* 31 (2003): 71–94.

Weinstein, Milton, George Torrance, and Alistair McGuire. "QALYs: The Basics." *Value in Health* 12, Supplement 1 (2009): S5–S9.

Weitzman, Gary. "Dr. Jane Goodall." In *The Animal House*. WAMU FM: American University Radio, September 19, 2009.

Wertheimer, Alan. *Coercion*. Princeton: Princeton University Press, 1987.

———. *Exploitation*. Princeton: Princeton University Press, 1996.

———. *Rethinking the Ethics of Clinical Research: Widening the Lens*. New York: Oxford University Press, 2010.

Whitener, Ron. "Research in Native American Communities in the Genetics Age." *Journal of Technology Law & Policy* 15 (2010): 217–74.

Wilkinson, Stephen. *Bodies for Sale*. London: Routledge, 2003.

Wood, Allen. *Kant's Ethical Thought*. Cambridge: Cambridge University Press, 1999.

——. *Kantian Ethics*. Cambridge: Cambridge University Press, 2008.

World Medical Association. "Declaration of Helsinki: Ethical Principles for Medical Research Involving Human Subjects." *Journal of the Indian Medical Association* 107 (2009): 403–5.

Zargooshi, Javaad. "Quality of Life of Iranian Kidney 'Donors.'" *The Journal of Urology* 166 (2001): 1790–9.

Index